Multiband Integrated Antennas for 4G Terminals

For a listing of recent titles in the *Artech House Antennas and Propagation Series*, turn to the back of this book.

Multiband Integrated Antennas for 4G Terminals

David A. Sánchez-Hernández
Universidad Politécnica de Cartagena

Editor

ARTECH HOUSE

BOSTON | LONDON
artechhouse.com

Library of Congress Cataloging-in-Publication Data
A catalog record for this book is available from the U.S. Library of Congress.

British Library Cataloguing in Publication Data
A catalogue record for this book is available from the British Library.

ISBN-13: 978-1-59693-331-6

Cover design by Yekaterina Ratner

© 2008 ARTECH HOUSE, INC.
685 Canton Street
Norwood, MA 02062

10 9 8 7 6 5 4 3 2 1

This book is dedicated to Matthew Steven, Daniel William, Katherine Mary, Gregory Reginald, Harriet, George, Isobel, Weiwei, Jowita Magdalena, Paulina Elzbieta, Jaroslaw Wojciech, Miguel, Lucía, Mario, Bruno, Irene, Marina, Emil, Søren, Mille, Lærke, Lykke and Liv, our current breed, and to Mary Caroline, Sarah, Shan Yan, Elzbieta, Ángel, José, José Manuel, Marien, Sylvia, Marina, Henriette and Ane, our mates. All of them had to spend less time with us so that we could write this book, and still remain unpaid for that.

Contents

Foreword

The worldwide communication technology thirst and demands in bringing digital information to widespread end users have revived the usage of small antennas on handsets and mobile units. Architectures of future wireless communication systems are gradually coming to the realization that antennas play a vital role in the optimal and effective performance of their mobile and personal units. It is now anticipated that the design of antennas and their implantations will be given more attention at the outset of the unit's design instead of waiting to the last minute. This is particularly becoming more critical due to the needs in operating in several frequency bands and at the same time occupying a very limited available area. Additionally, constant demands in increasing information capacity have attracted utilization of multiple antennas either in diversity or MIMO operational modalities.

This book attempts to address some fundamental issues concerning small antenna designs and provide relevant representative recent developments. The seven chapters of the book are written by international experts from academia and industry who present a balanced approach among various important technical topics pertinent to small antennas. Clearly there is no single antenna design that addresses all the needs of various users and diverse variations in mobile unit's architectures and form factors. Nevertheless, there are ample recent developments discussed in this book that should provide useful guidelines for the antenna designers working in these areas.

One thing is clear that the system engineers need to lower their expectations in terms of realistic operational performance of these small antennas. There is no room for a large ground plane, there are interaction issues with other components, there is interaction with the user, and so forth. Therefore, one should not expect to achieve −10-dB return loss in the entire frequency bands of interests nor one can expect pure polarization performance and clean radiation patterns. In some applications depolarization could help the overall unit performance.

In the past, one antenna operating at one frequency band was typically the design objective. Today, there are several frequency bands and multiple antennas are required to respond to the needed information capacity and complex environmental and propagation manifests. Fortunately, advanced and powerful simulation tools have allowed the antenna researches and designers to explore the possibilities of utilizing designs that were not even dreamed of in the rather recent past. Advanced global optimization techniques have also helped in this endeavor; however, there still is a need to understand the underlying physics of this class of antennas. Various chapters of this edited book should provide a good starting point for practitioners and researches in these evolving and complex design paradigms.

I strongly believe that we are in the era that antennas for wireless communications applications are manufactured in billions. Moreover, we must remember that at no time in history have antennas come so close to the human. It is our responsibility to design effective, power-efficient, and radiation-safe devices. Antenna engineers play an important role in this digital information technology age and will influence its successful future!

Yahya Rahmat-Samii
Distinguished Professor
Electrical Engineering Department
University of California, Los Angeles
July 2008

Preface

An enormous amount of research papers are available in the scientific literature for MIMO systems. Despite initial quasi-utopist ergodic capacities being considerably reduced by real implementation prototypes, MIMO techniques are foreseen as the key for solving current capacity and coverage problems of mobile communications systems. 3G bottlenecks have already been identified, and the highly efficient proposed 4G technologies will certainly make use of MIMO techniques to further reduce interference and provide for the extra capacity required in future applications, particularly at downlink. However, this will most probably require multiple antennas at both ends of the link, which highlights the importance of handset MIMO. Unlike his elder conventional MIMO brother, handset MIMO has not received as much attention. The challenges for the antenna designer are certainly astonishing in handset MIMO. While in current mobile handset designs less and less volume is made available for the antenna, at the same time this radiating element has to be able to operate on an increasingly larger number of frequency bands, corresponding to the diverse number of wireless systems typically combined in a handset. Miniaturization, integration, multiband operation, and final performance were real challenges in early designs. A few years ago acquiring the bandwidth requirements for multiband handsets with relatively relaxed VSWR and integration within the handset case was just about the only petition to handset antenna designers. Extraordinarily quickly developed multiband techniques, some of which will be described in this book, soon made multiband handsets a commercial reality. In addition, computing capabilities of handset on-board processors, mainly aimed at the need for real-time video monitoring, vectorial-game-processing, or functional wideband access to Internet, have increased considerably due to market demands. This has conveniently helped space-time processing (STP) algorithm development, which is currently being beta tested for handsets. Surprisingly, smart MIMO multiband handset antenna engineering has not received as much attention, although this tendency is rapidly changing. While antenna designers are ready for new challenges, the need for handset MIMO is currently market-driven, but information is scarce and rarely shared at commercial level. In consequence, the aim of this book is to provide the advance reader with a conjunction of the recently developed multiband techniques for handset antennas and their inherent possibilities and limitations in handset MIMO, conveniently jointly studied and congruently united in a single book by several experts in the field.

The book begins with an introductory chapter (Chapter 1) covering multiband antenna theory and size reduction techniques. There are specific issues to be addressed when reducing the size of an antenna and the performance of small antennas and small antenna theory will help the reader understand the rest of the book. Chapter 2 is devoted to antenna designs capable of both multiband and multisystem operation. While wideband is a commonplace need for fixed wireless communications, the requirements for multiband integrated antennas may have relaxed a little in comparison. VSWR or radiation patterns may not need to be as good as when the antenna is not integrated. However, multiband techniques making use of the available volume are not simple, and are inherently dressed up with some drawbacks that will be highlighted in this chapter. Among the diverse multiband techniques, there has been some discussion about the trade-offs of using wirelike or patchlike antennas for handset antenna design. In Chapter 3, planar wire antennas will highlight the benefits of a hybrid solution. In contrast, Chapter 4 will concentrate on the benefits of printed fractal techniques on handset antenna design. Particular emphasis is placed on the multiband properties and capabilities of these antennas. PIFA-like slotted patch antennas will be reviewed in Chapter 5, as well as some innate practical aspects of miniaturization, such as the effect of the chassis, on final performance. With all previous multiband antenna designs in mind, the antenna engineer faces the further challenges of the future advent of 4G. MIMO techniques are a hot research line since they are part of 4G systems, but the specific aspects of MIMO techniques when employed in a reduced-volume such as the handset are yet to be fully addressed. Chapter 6 is aided to identify the effects of radiation and mismatch efficiency, inherently critical when antennas are integrated in a handset, on final MIMO capacity or diversity gain figures. This includes the effect of the presence of the user on final handset MIMO performance. Likewise, novel MIMO techniques to be specifically ideal for handset design, such as the true polarization diversity (TPD), are explained and compared to more conventional spatial, pattern, or polarization diversity techniques. Finally, Chapter 7 is aimed to identify the current role of antenna design in 4G handset production when evaluated in terms of communications performance, and it also serves as the conclusion. Chapter 7 reviews the most important parameters when evaluating communications performance and highlights the role of the antenna system and its effect on these parameters. This last chapter is very helpful for identifying key design parameters since the impact on final performance results is deeply analyzed. The importance of an adequate antenna system design and construction is clearly reinforced in this chapter.

With the information provided in all chapters it becomes obvious that the antenna design has to be integrated into handset design at the very early stages, and that an increasingly important role of antenna design will require hands-on knowledge of both theoretical and practical aspects. While tremendous efforts are placed on gaining 1 or 2 dB in the MMIC, 10-dB differences in mean effective gain (MEG) are commonplace, and the use of smartly designed and truly integrated antennas could help the communications performance much more that it does today.

I sincerely hope that you find this book useful for your research or technological projects, and apologize in advance for the unavoidably complex scientific jargon and structure of the material we have prepared, as well as for those inevitable mistakes we could have made in the writing and editing processes.

I also have to express my sincere thanks to all authors for accepting my invitation, for their patience on my sticking-to-schedule demands, and particularly to Eric Willner, Jessica Thomas, Allan Rose, Rebecca Allendorf, and Barbara Lövenvirth, commissioning and developmental editors for the publisher at different stages, for their encouragement and support all throughout the making of this work.

Good reading.

David A. Sánchez-Hernández
Editor
Technical University of Cartagena
Cartagena, Spain
July 2008

CHAPTER 1

Electrically Small Multiband Antennas

Steven R. Best

1.1 Introduction

In today's environment of almost constant connectivity, wireless devices are ubiquitous. As the RF electronics technology for these wireless devices continues to decrease in size, there is a corresponding demand for a similar decrease in size for the antenna element. Unfortunately, the performance requirements for the antenna are rarely relaxed with the demand for smaller size. In fact, the performance requirements generally become more complex and more difficult to achieve as the wireless communications infrastructure evolves.

In the early deployment of cellular, Digital Cellular System (DCS), Personal Communications Service/System (PCS), and Global System for Mobile Communications (GSM) networks, the wireless device typically had to operate within a single band, defined by the specific carrier's license(s). In today's environment, the wireless device is often required to operate in more than one band that may include several GSM frequencies, 802.11 (Wi-Fi), 802.16 (Wi-Max) and Global Positioning System (GPS) as defined in Table 1.1. Frequencies below GSM 800 are not presented.

From the antenna engineer's perspective, an antenna that operates continuously without tuning from 824 MHz through 2,500 MHz is considered a single band but also wideband antenna. In today's wireless device terminology, an antenna that covers more than one of the wireless communications bands is considered a multiband antenna. For example, an antenna that simultaneously covers two separate bands encompassing frequencies of 824–960 MHz and 1,710–1,990 MHz is considered a four-band or quad-band antenna since it provides coverage of the GSM 800, GSM 900, GSM 1800, and GSM 1900 frequencies.

In addition to dealing with the challenges of designing a small antenna, antenna engineers must also deal with the challenge of designing the small antenna to operate over multiple frequency bands. The chapters that follow in this book discuss many of the advanced techniques used in the design of small multiband antennas. This chapter serves as an introduction to the remaining chapters in that it introduces the definition of a small antenna, it provides definitions of the fundamental performance properties of antennas, and it details fundamental limitations associated with the design of small antennas. In addition, this chapter provides a discussion on the most fundamental approaches and antenna types used in the design of small wideband antennas. Finally, it concludes with a discussion on the effects of the finite ground plane on antenna performance, a significant issue in the design of integrated antennas, where the ground plane may be a significant or dominant portion of the radiating structure.

Table 1.1 Wireless Communications Bands and Their Frequency Designations

Band Designation	Alternate Description(s)	Transmit Frequency (Uplink) (MHz)	Receive Frequency (Downlink) (MHz)
GSM 800 or GSM 850	AMPS DAMPS	824–849	869–894
P-GSM 900	Primary GSM 900	890–915	935–960
E-GSM 900	Extended GSM 900	880–915	925–960
GSM-R 900	Railways GSM 900	876–915	921–960
T-GSM 900	TETRA GSM 900	870.4–915	915.4–921
GPS		N/A	1,565.42–1,585.42
GSM 1800	DCS 1800	1,710–1785	1,805–1880
GSM 1900	PCS 1900	1,850–1910	1,930–1990
UMTS		1,885–2,025 1,710–1,755 (US)	2,110–2,200 2,110–2,155 (US)
802.11 b/g/n	Wi-Fi; ISM	2.4 – 2.4835 GHz ISM	
802.11 a/h/j	Wi-Fi; UNII	5.15–5.35 GHz (UNII) 5.47–5.725 GHz 5.725–5.825 GHz (ISM/UNII) 4.9–5 GHz (Japan) 5.03–5.091 GHz (Japan)	
802.15.4	Zigbee	898 MHz 915 MHz 2.4-GHz ISM	
802.15.1 1a	Bluetooth	2.4–2.4835-GHz ISM	
802.15.3	UWB	Typically > 500 MHz bands within the 3.1 – 10.6-GHz spectrum	
802.16	Wi-Max	Various bands within the 2–11-GHz spectrum. Mobile Networks: 2–6 GHz Fixed Network: < 11 GHz	

1.2 The Definition of Electrically Small

Prior to defining the performance properties and fundamental limitations of the electrically small antenna, it is necessary to define the parameter that establishes whether or not the antenna is in fact, electrically small. Consider a typical device integrated antenna mounted at the center of a large, conducting ground plane, as shown in Figure 1.1. Mounted at the center of a large ground plane, the

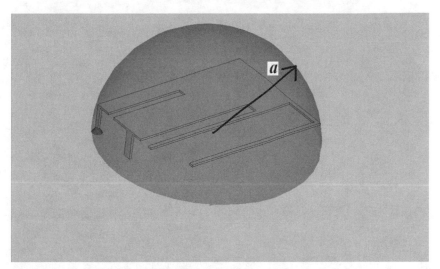

Figure 1.1   Depiction of the *ka* sphere for an antenna that is mounted at the center of a large ground plane. The radius, *a*, of the *ka* sphere encompasses the maximum dimension of the antenna.

antenna's impedance and bandwidth properties are similar to those where the antenna is mounted at the center of an infinite ground plane. In this case, the antenna is considered to be electrically small when the value of $ka \leq 0.5$, where k is the free space wavenumber $2\pi/\lambda$, where λ is the free space wavelength, and a is the radius of an imaginary sphere circumscribing the maximum dimension of the antenna as shown in Figure 1.1. If the antenna were a vertical, straight-wire monopole, a would be equal to the monopole's height. As shown for the antenna depicted in Figure 1.1, the value of a must encompass both the vertical and horizontal dimensions of the antenna.

If the small device antenna is operated over a small ground plane, as is almost always the case, the current distribution on the ground plane often becomes a significant or dominant portion of the antenna structure. In this instance, the impedance and bandwidth properties of the antenna may be substantially different than when the antenna is mounted at the center of a large ground plane. For these antennas, the entire ground plane structure must be included in the definition of a, as illustrated in Figure 1.2.

1.3   Fundamental Antenna Properties Definitions

The performance of single and multiband antennas is characterized by a number of electrical properties, which include impedance, VSWR, bandwidth, gain (a function of directivity, efficiency, and mismatch loss), and polarization. Definitions of these properties are well known but are included here for completeness.

1.3.1   Impedance, Efficiency, VSWR, and Gain

The complex, frequency-dependent input impedance of the small antenna is given by [1]

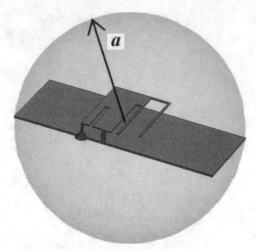

Figure 1.2   Depiction of the *ka* sphere for an antenna that is mounted on a small device-sized ground plane. The radius, *a*, of the *ka* sphere encompasses the maximum dimensions of both the ground plane and antenna.

$$Z_A(\omega) = R_A(\omega) + jX_A(\omega) \tag{1.1}$$

where $R_A(\omega)$ and $X_A(\omega)$ are the antenna's frequency-dependent resistance and reactance, respectively, and ω is the radian frequency, $2\pi f$, where f is the frequency in hertz.

If the small antenna behaves as a series RLC circuit, the frequency derivative of its reactance, $X_A{}'(\omega)$, will be greater than zero at its first resonance (a series resonance), where $X(\omega_0) = 0$ and ω_0 is the resonant frequency. When the small antenna behaves as a series RLC circuit, its reactance will approach $-\infty$ as $\omega \to 0$. At very low frequencies, the antenna has the impedance properties of a lossy capacitor. In this case, the antenna's loss resistance, $R_l(\omega)$ is in series with its radiation resistance, $R_r(\omega)$, and $R_A(\omega) = R_r(\omega) + R_l(\omega)$. The antenna's frequency dependent radiation efficiency is given by

$$\eta_r(\omega) = \frac{R_r(\omega)}{R_A(\omega)} = \frac{R_r(\omega)}{R_r(\omega) + R_l(\omega)} \tag{1.2}$$

For the small antenna that behaves as a series RLC circuit, the antenna's radiation resistance is given by $R_r(\omega) = \eta_r(\omega) R_A(\omega)$ and the antenna's loss resistance is given by $R_l(\omega) = [1 - \eta_r(\omega)] R_A(\omega)$.

If the small antenna behaves as a parallel RLC circuit, the frequency derivative of its reactance, $X_A{}'(\omega)$, will be less than zero at its first resonance (a parallel resonance or antiresonance). When the small antenna behaves as a parallel RLC circuit, its reactance will approach 0 as $\omega \to 0$. At very low frequencies, the antenna has the impedance properties of a lossy inductor. In frequency regions very near the antiresonance, the antenna's loss resistance, $R_l(\omega)$, is in parallel with its radiation resistance, $R_r(\omega)$, and

$$R_A(\omega) = \frac{R_r(\omega)R_l(\omega)}{R_r(\omega) + R_l(\omega)} \qquad (1.3)$$

The antenna's frequency-dependent radiation efficiency is given by

$$\eta_r(\omega) = \frac{R_A(\omega)}{R_r(\omega)} = \frac{R_l(\omega)}{R_r(\omega) + R_l(\omega)} \qquad (1.4)$$

For the small antenna that behaves as a parallel RLC circuit, the antenna's radiation resistance is given by $R_r(\omega) = R_A(\omega)/\eta_r(\omega)$ and the antenna's loss resistance is given by $R_l(\omega) = R_A(\omega)/[1 - \eta_r(\omega)]$. Well away from the antiresonance ($\omega \to 0$), the radiation and loss resistances will be in series, similar to those of the small antenna that behaves as a series RLC circuit.

Any small antenna that behaves as a series RLC circuit is dipole- or monopole-like from an impedance perspective and can be identified by the fact that there is not a DC short anywhere within the antenna structure that connects the center conductor of the feed cable to its shield or ground. Any small antenna that behaves as a parallel RLC circuit is looplike from an impedance perspective and can be identified by the fact that there is a DC short somewhere within the antenna structure that connects the center conductor of the feed cable to its shield or ground. Dipole- and monopole-like antennas behave as a series RLC circuit, where their radiation and loss resistances are in series from very small values of ω through their first series resonant frequency. Looplike antennas behave as a series RLC circuit for very small values of ω ($ka \ll 0.5$). However, as ω approaches the loop's first resonant frequency (an antiresonance), the antenna behaves as a parallel RLC circuit and the radiation and loss resistances are in parallel.

For all antennas, the input voltage standing-wave ratio (VSWR, designated s) is given by

$$VSWR(\omega) = s = \frac{1 + |\Gamma(\omega)|}{1 - |\Gamma(\omega)|} \qquad (1.5)$$

where $\Gamma(\omega)$ is the frequency-dependent complex reflection coefficient given by

$$\Gamma(\omega) = \frac{Z_A(\omega) - Z_{CH}}{Z_A(\omega) + Z_{CH}} \qquad (1.6)$$

where Z_{CH} is the real (not complex) characteristic impedance of the transmission line connecting the antenna to a transmitter. While VSWR provides a quantitative indication of the level of mismatch between the antenna's impedance and the transmission line's characteristic impedance, a more direct measure relating the amount of power reflected at the antenna's feed point terminal to the incident or forward power is the mismatch loss (ML) defined by

$$ML(\omega) = 1 - |\Gamma(\omega)|^2 \qquad (1.7)$$

In practice, the VSWR and mismatch loss are often characterized by the return loss (RL), which is defined as 20log(|Γ|). For most wireless, integrated device antennas, the VSWR is required to be less than 3 throughout the designated operating band(s). This is equivalent to a return loss requirement of 6 dB.

If the antenna's pattern directivity is defined D, then the antenna's overall or realized gain is defined

$$G(\omega) = \eta_r(\omega)[1 - |\Gamma(\omega)|^2]D \qquad (1.8)$$

With decreasing value of ka below 0.5, the directivity of a dipole approaches a constant of 1.5 (approximately 1.8 dB), while the directivity of monopole on an infinite, highly conducting ground plane approaches a constant of 3 (approximately 4.8 dB). If the small dipole and monopole are perfectly impedance matched and lossless, their realized gains approach theoretical maximum values of 1.8 and 4.8 dBi, respectively. In the case of the device integrated monopole (or other small device integrated antenna), the ground plane is small in extent and may have dimensions approaching or less than the value of the operating wavelength. In this case, the directivity is typically less than 3 (4.8 dB) and the maximum realized gain of the antenna is less than 4.8 dBi. The typical value of realized gain for efficient antennas ranges between 2 and 4 dBi.

1.3.2 Bandwidth and Quality Factor (Q)

One of the challenges associated with the design of device integrated antennas is achieving the desired VSWR over one or multiple operating bands. Operating bands of interest may include any number of the usual communication bands defined in the introduction of this chapter. From a practical perspective, the engineer is usually interested in the VSWR bandwidth, defined by a specific value of VSWR, s, relative to the characteristic impedance of a transmission line (typically 50Ω) or the output impedance of a transmit amplifier or receive circuit. This VSWR bandwidth can easily be determined through simulation or measurement.

Beyond this, we are also interested in definitions of bandwidth and quality factor (Q) that allow the engineer to determine how well the antenna design performs relative to theoretical limits. For a small antenna of a given size (ka), there are established, fundamental limits for the minimum value of Q that can be achieved. Often, the definition of bandwidth can be ambiguous; however, here we define bandwidth two specific ways.

For the antenna that exhibits a single impedance resonance within its defined VSWR bandwidth as shown in Figure 1.3, and where the bandwidth is relatively small (a narrowband antenna), we define the VSWR bandwidth in terms of the fractional bandwidth as [2]

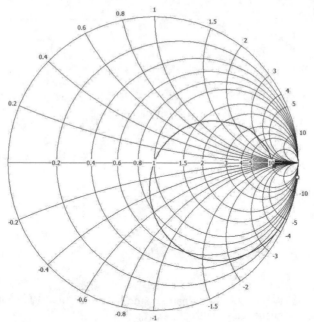

Figure 1.3   Typical feed-point impedance of an antenna that exhibits a single impedance resonance within its defined VSWR bandwidth.

$$FBW = \frac{f_h - f_l}{f_0} \qquad (1.9)$$

where f_h and f_l are the frequencies for an arbitrary VSWR, s, above and below, the frequency of minimum VSWR, f_0, respectively, as shown in Figure 1.4. For narrowband antennas, the value of fractional bandwidth, FBW, is assumed to be less than approximately 25%.

For the antenna that exhibits multiple impedance resonances or a single wideband resonance within its defined VSWR bandwidth as shown in Figure 1.5, and where the bandwidth is relatively large (a wideband antenna), we define the VSWR bandwidth in terms of the bandwidth ratio as

$$BWR = \frac{f_h}{f_l} \qquad (1.10)$$

where f_h and f_l are the upper and lower frequencies defining the full extent of the operating band for a defined VSWR, s, as shown in Figure 1.6. For wideband antennas, the value of the bandwidth ratio, BWR, is assumed to be greater than approximately 1.25.

It is important to note here again the distinction in the definition of bandwidth for many device integrated antennas. The antenna may have one region of VSWR bandwidth that simultaneously covers multiple communications bands; that is, an antenna with a VSWR less than 2 from 1,710 through 1,990 MHz covers two GSM bands: GSM 1800 (1,710–1,785 MHz; 1,805–1,880 MHz), and GSM 1900

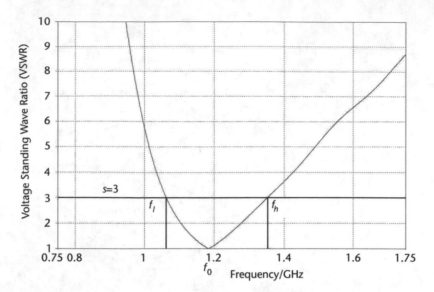

Figure 1.4   VSWR bandwidth for narrowband antenna. The VSWR bandwidth is defined by the fractional bandwidth, $(f_h - f_l)/f_0$.

(1,850–1,910 MHz; 1,930–1,990 MHz). In this case, the antenna engineer may consider the antenna a single-band antenna; however, industry standards would typically consider this antenna a dual-band antenna because it covers two communications bands.

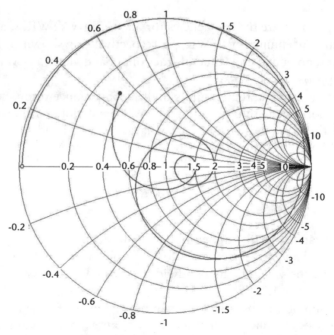

Figure 1.5   Typical feed-point impedance of an antenna that exhibits multiple impedance resonances within its defined VSWR bandwidth.

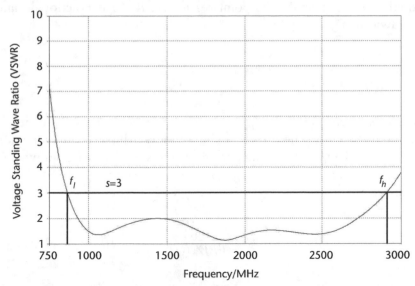

Figure 1.6   VSWR bandwidth for a wideband antenna. The VSWR bandwidth is defined by the bandwidth ratio, f_h/f_l.

Unlike bandwidth, the definition for the Q of an antenna is not ambiguous. There is only one definition for exact Q, which is defined for the resonant or tuned antenna as [3],

$$Q(\omega_0) = \frac{\omega_0 W(\omega_0)}{P_A(\omega_0)} \qquad (1.11)$$

where ω_0 is the resonant or tuned frequency, $W(\omega_0)$ is internal energy, and $P_A(\omega_0)$ is the power accepted by the antenna, including the powers associated with radiation and loss. Rather than using the exact definition of Q as stated in (1.11), it is more appropriate to use a definition of Q that is stated in terms of the antenna's impedance properties and bandwidth. This was done in [2], where it was shown that the exact Q of the antenna can be approximated as

$$Q(\omega_0) \approx Q_Z(\omega_0) = \frac{\omega_0}{2R_A(\omega_0)} \sqrt{ R_A'(\omega_0)^2 + \left(X_A'(\omega_0) + \frac{|X_A(\omega_0)|}{\omega_0} \right)^2 } \qquad (1.12)$$

where $R_A'(\omega)$ is the frequency derivative of the antenna's resistance. In (1.12), the frequency derivatives are taken with respect to ω, not f. Having defined antenna Q, it is now necessary to define a bandwidth where Q and the inverse of bandwidth are related over all ranges of frequency. The definition of bandwidth suitable for this purpose is fractional VSWR bandwidth with the specific distinction that fractional matched VSWR bandwidth is used. In this case, the fractional matched VSWR bandwidth, $FBW_V(\omega_0)$, is defined where the VSWR of the tuned or self-resonant antenna is determined using a characteristic impedance, Z_{CH}, equal to the antenna's feed-point resistance, $R_A(\omega_0)$. Fractional matched VSWR bandwidth is given by [2]

$$FBW_V(\omega_0) = \frac{\omega_+ - \omega_-}{\omega_0} = \frac{f_h - f_l}{f_0} \tag{1.13}$$

where ω_+ $(2\pi f_h)$ and ω_- $(2\pi f_l)$ are the frequencies above and below ω_0 $(2\pi f_0)$, respectively, where the VSWR is equal to any arbitrary value denoted by s. Note that in (1.13), f_0 is the frequency where the $s = 1$, rather than where it is the frequency of minimum VSWR as in (1.9). Fractional matched VSWR bandwidth and Q are related as [2]

$$Q(\omega_0) \approx Q_B(\omega_0) = \frac{2\sqrt{\beta}}{FBW_V(\omega_0)}, \qquad \sqrt{\beta} = \frac{s-1}{2\sqrt{s}} \le 1 \tag{1.14}$$

The approximations in (1.12) and (1.14) were derived in [2] under the assumptions that the tuned or self-resonant antenna exhibits a single impedance resonance within its defined operating bandwidth and that the half-power matched VSWR bandwidth is not too large.

1.4 Fundamental Limits of Small Antennas

In the previous section, basic definitions for the performance properties of small antennas were presented. These definitions included impedance, radiation efficiency, VSWR, mismatch loss, realized gain, bandwidth, and quality factor. The question addressed in this section is what are the fundamental limits to these performance properties as a function of antenna size or volume, ka?

Any small antenna can be impedance-matched at any single frequency using a number of techniques that can include an external matching network or internal modification of the antenna structure. While impedance matching the antenna mitigates mismatch loss, there are ohmic loses associated with the specific implementation of a matching network. These losses reduce the radiation efficiency of the antenna, but ideally they are less than the losses associated with the initial mismatch between the antenna and the connecting transmission line or transmitter/receiver circuit. In theory, there is no upper bound on the radiation efficiency of the small antenna, so there is no theoretical upper bound on the realized gain, except that it cannot exceed the antenna's directivity. While there is a theoretical upper bound on the small antenna's directivity, this is not specifically addressed here since it assumed that the small antenna's directivity is limited by the small dipole and monopole directivities of 1.8 and 4.8 dB, respectively.

From a practical perspective, the realized gain is limited by issues such as minimizing the losses within any impedance matching network, achieving maximum radiation efficiency with appropriate choices of conductor dimensions and dielectric materials, and minimizing levels of radiation into unintended polarization. In many wireless communications systems where the communications link may be established through multipath, some randomness in the antenna's polarization properties may be desirable.

It is also important to note that some issues associated with achieving a specific realized gain do not scale with frequency while maintaining the same value of *ka*. First, conductor and dielectric losses do not scale with frequency. For example, an antenna designed to operate at a low frequency will have a higher efficiency than it will when it is exactly scaled (all dimensions) to operate at a higher frequency. Second, practical issues with implementing an antenna can change dramatically when changing frequency. For example, it is physically easier to implement an antenna having $ka = 0.1$ at a frequency of 1 MHz, where $a/\lambda = 4.77$ m, than it is at 1 GHz, where $a/\lambda = 4.77$ mm.

While there is no theoretical limit on the small antenna's realized gain (other than the conditions stated above), there is a theoretical limit on the minimum achievable Q as a function of its size relative to the operating wavelength (ka). Since Q and bandwidth are inversely related (in a single resonance antenna), there is a corresponding upper limit on the maximum achievable bandwidth. The work of Chu [3] and McLean [4] has led to the commonly accepted lower bound on Q (often referred to as the Chu limit) being defined as

$$Q_{lb} = \eta_r \left(\frac{1}{(ka)^3} + \frac{1}{ka} \right) \qquad (1.15)$$

The radiation efficiency term in (1.16) indicates that lossy antennas can achieve a lower Q, and hence increased bandwidth, relative to their lossless counterparts. Substituting (1.15) into (1.14), an upper bound on the fractional matched VSWR bandwidth can be written as

$$FBW_{Vub} = \frac{1}{\eta_r} \frac{(ka)^3}{1 + (ka)^2} \frac{s-1}{\sqrt{s}} \qquad (1.16)$$

A plot of the upper bound on fractional matched VSWR bandwidth versus *ka* for $s = 3$ is presented in Figure 1.7. This upper bound is for the lossless antenna. At $ka = 0.5$, the upper bound on fractional matched VSWR bandwidth is approximately 11.5%. At $ka = 0.1$, the upper bound decreases to approximately 0.11%, illustrating the significant bandwidth limitation that occurs when the size of the antenna decreases relative to the operating wavelength.

The upper bound on bandwidth defined in (1.15) is derived under the assumptions that the antenna exhibits a single resonance within its operating bandwidth, it is perfectly matched at the resonant frequency, and there is no external matching network. In the design of most antennas, the antenna is not perfectly matched at its resonant frequency or at its minimum value of VSWR, which may not correspond to a resonant frequency. Typically, a slight increase in bandwidth can be achieved if the antenna is slightly mismatched and does not exhibit a minimum VSWR, $s = 1$.

Additionally, an external matching network can be used to increase the bandwidth of the small antenna, with the corresponding trade-offs in both gain (efficiency) and bandwidth. Bandwidth matching limits (Fano limits) exist when

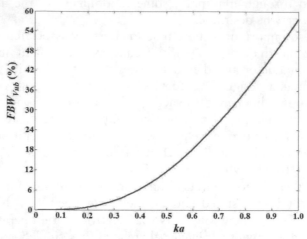

Figure 1.7   Upper bound on fractional matched VSWR bandwidth as a function of *ka* for a VSWR, *s* = 3.

externally impedance matching an antenna of a given Q using an N-section, reactive matching network [5].

For an infinite number of external impedance matching sections, the maximum achievable bandwidth is given by [6]

$$BW_\infty = \frac{1}{Q} \frac{\pi}{\ln\left(\frac{1}{|\Gamma|}\right)} \tag{1.17}$$

Substituting the lower bound on Q from (1.15) into (1.17), and expressing $|\Gamma|$ in terms of an arbitrary VSWR, *s*, the maximum fractional matched VSWR bandwidth that can be achieved with an infinite number of matching sections is

$$FBW_{V\max} = \frac{1}{\eta_r} \frac{(ka)^3}{1+(ka)^2} \frac{\pi}{\ln\left(\frac{s+1}{s-1}\right)} \approx \frac{1}{\eta_r} \frac{\pi(ka)^3}{\ln\left(\frac{s+1}{s-1}\right)}; \quad (ka << 1) \tag{1.18}$$

For the half-power VSWR bandwidth (*s* = 5.828), this is approximately 4.5 times the upper bound on fractional bandwidth of (1.16) that can be achieved without external matching circuits for the same antenna Q.

1.5   Special Considerations for Integrated Multiband Antennas

In the previous section, the fundamental limitations on the quality factor and bandwidth of the small antenna as a function of *ka* were presented. In this discussion, it was assumed that the small antenna exhibits a single impedance resonance within its defined VSWR bandwidth, as was illustrated previously in Figure 1.3. When there is only a single impedance resonance within the antenna's defined

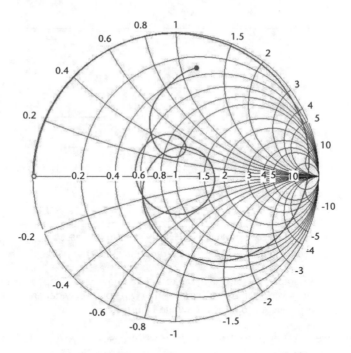

Figure 1.8   Impedance of a wideband antenna with closely spaced multiple impedance resonances within its defined VSWR bandwidth.

bandwidth, the frequency derivatives at the resonant frequency provide a very
good approximation of the impedance properties at the band edges and the
approximation of antenna Q in (1.12) holds.

When there are closely spaced multiple impedance resonances within the
antenna's defined operating bandwidth, as shown in Figure 1.8, the frequency
derivatives at any one resonance do not necessarily provide a good approximation of the impedance properties at the band edges and the approximation of
antenna Q in (1.12) may not hold. Additionally, in some instances, the inverse
relationship between bandwidth and Q, as defined in (1.14), may not hold, particularly as the value of s increases [7, 8]. This is illustrated in Figure 1.9, where
the values of Q_Z, determined from (1.12) are compared with the values of Q_B,
determined from (1.14). The quality factor in Figure 1.9 corresponds to the
impedance data in Figure 1.8.

At very low frequencies, there is excellent agreement between the values of Q_Z
and Q_B. With increasing frequency, and in the regions where the antenna exhibits
closely spaced resonances, the agreement between Q_Z and Q_B degrades, particularly with increasing values of s. For $s = 5.828$ (half-power), the agreement is poor
over most of the operating band. Notice as well that the equivalent Q determined
from the inverse of bandwidth (Q_B) more closely approaches the lower bound
than does the equivalent Q determined from the impedance derivatives (Q_Z).
While it may be possible to design an antenna whose inverse bandwidth is less
than the lower bound, the antenna's exact Q, precisely determined from (1.11)
will never be less than the lower bound of (1.15).

As is typically done in the design of many multiband antennas, consider the
return loss properties of three antennas as illustrated in Figures 1.10, 1.11, and

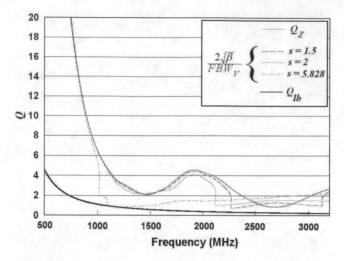

Figure 1.9   A comparison of quality factors determined from impedance derivatives (Q_Z) and the inverse of exact bandwidth (Q_B) for the wideband antenna exhibiting the impedance properties shown in Figure 1.8.

1.12. Figure 1.10 depicts the return loss of a single resonant (band) antenna, Figure 1.11 depicts the return loss of an antenna exhibiting multiple operating bands spaced sufficiently apart such that the impedance is not well matched between resonances, and Figure 1.12 depicts the return loss of an antenna with two operating bands where the impedance between bands is well matched.

For an antenna exhibiting the return loss behavior illustrated in Figure 1.10, there is a single impedance resonance (or antiresonance) within the defined VSWR bandwidth and the relationships between impedance, bandwidth, and Q defined in (1.12) and (1.14) will hold over a wide range of frequencies. Additionally, the

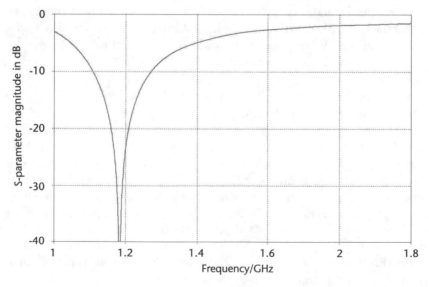

Figure 1.10   Return loss of an antenna exhibiting a single resonance within its defined VSWR bandwidth, and a single operating band over a given range of frequencies.

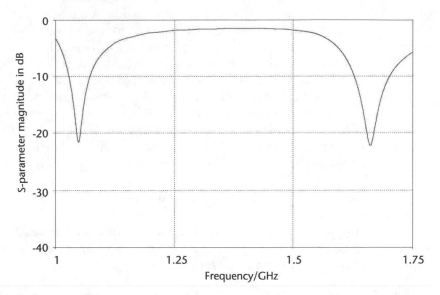

Figure 1.11   Return loss of an antenna exhibiting a multiple resonances or operating bands that are spaced such that the impedance between the operating bands is not well matched.

equivalent Q determined from the inverse of the antenna's bandwidth, Q_B, will be greater than the lower bound defined in (1.15).

For an antenna exhibiting the return loss behavior illustrated in Figure 1.11, there will typically be a single impedance resonance (or antiresonance) within each of the two operating bands, which are separated by an impedance antiresonance (or resonance) that is well outside the VSWR specification. In each of these matched impedance bands, the relationships between impedance, bandwidth, and

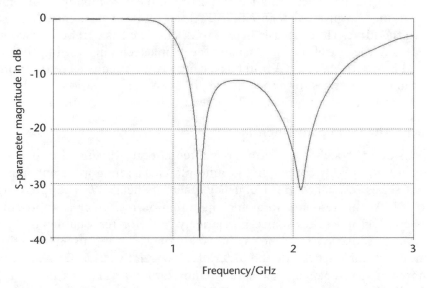

Figure 1.12   Return loss of an antenna exhibiting multiple resonances or operating bands that are spaced such that the impedance between the operating bands is well matched, forming a much wider operating bandwidth.

Q defined in (1.12) and (1.14) will hold over a wide range of frequencies. Additionally, the equivalent Q determined from the inverse of the antenna's bandwidth, Q_B, will be greater than the lower bound defined in (1.15).

For the antenna exhibiting the return loss behavior depicted in Figure 1.12, where the impedance match extends over multiple operating bands or frequencies of minimum return loss, there will be multiple impedance resonances (and/or antiresonances) within the specified VSWR bandwidth and the relationships between impedance, bandwidth, and Q, defined in (1.12) and (1.14) may not hold over narrow or wide ranges of frequency. Additionally, the equivalent Q determined from the antenna's impedance properties, Q_Z, and the inverse of the antenna's bandwidth, Q_B, may be less than the lower bound defined in (1.15). However, the antenna's exact Q determined from (1.11) will always be greater than this lower bound.

Finally, in any of the return loss graphs presented in Figures 1.10 through 1.12, it is possible for closely spaced multiple impedance resonances to exist within the frequencies of minimum return loss.

1.6   Fundamental Approaches to Small Integrated Antenna Design

The primary focus of this text is the discussion of the design and performance properties of multiband integrated antennas. While this chapter covers introductory or fundamental topics, the remainder of the book focuses on general and specific design approaches for achieving multiband operation with an integrated antenna. These approaches include discussions and examples of planar (wire and patch) antennas, multiband fractal antennas, the planar inverted-F antenna (PIFA), and so forth.

Prior to the specific details of these advanced design approaches, this section discusses the basic performance properties of the wire monopole antenna, its transition into an inverted-L antenna (ILA), an inverted-F antenna (IFA), and finally into the PIFA. The emphasis here is to discuss the basic trade-offs associated with reducing the size of the antenna and the simple technique for impedance matching used in the IFA and PIFA antennas. There is no attempt here to optimize any of the antenna designs for a specific size, frequency of operation, bandwidth, or multiple band operation, which will be dealt with in other chapters.

1.6.1   The Monopole Antenna

Consider a straight-wire monopole antenna that is assumed to be mounted over an infinite, highly conducting ground plane. The monopole antenna exhibits its first series resonance at a frequency where its height, h, is approximately equal to $\lambda/4$. At this resonant frequency its radiation resistance is approximately 36Ω, its conductor loss resistance is relatively low (for reasonable conductor diameters), its radiation efficiency is in excess of 90%, and its 3:1 fractional VSWR bandwidth is typically on the order of 25% (relative to 50Ω, the monopole is not matched at resonance). The actual bandwidth will vary as a function of conductor diameter.

In virtually all modern wireless communication applications in the 824-MHz–2.5-GHz frequency range, the $\lambda/4$ monopole (and associated ground plane) is too large for device integration. With decreasing frequency (and fixed height) or decreasing height (and fixed frequency), the monopole's reactance becomes increasingly capacitive and the radiation resistance is eventually ($ka \leq 0.5$) given by [1],

$$R_r \approx 40\pi^2 (\frac{h}{\lambda})^2 \tag{1.19}$$

The increase in capacitive reactance and reduction in radiation resistance result in an increase in VSWR, mismatch loss, and a reduction in radiation efficiency. There is also an associated decrease in the fractional VSWR bandwidth with decreasing height. Both the decrease in radiation resistance and bandwidth have been shown to be directly a function of the reduction in antenna height, or more precisely, the value of $(h/\lambda)^2$ [9]. Typically, in the design of small antennas, the first significant challenge is mitigating the high VSWR and mismatch loss by first tuning the antenna to resonance and then impedance matching the antenna to the desired characteristic impedance Z_{CH}, or designing the antenna such that its radiation resistance is equal to Z_{CH}.

For reasonable values of ka, ($0.1 \leq ka \leq 0.5$), the radiation efficiency may be less of an issue with appropriate choice of conductor diameter and antenna design. In addition to being a function of ka, practical issues associated with the choice of design frequency also become significant. As noted earlier, it is easier to design a small antenna at HF frequencies than it is at the frequencies associated with typical wireless communications systems, which primarily operate between 800 MHz and 2.5 GHz or higher.

Finally, the maximum achievable bandwidth is generally limited by the lower and upper bounds defined in previous sections. Achieving bandwidths that approach these bounds can be challenging and often requires judicious approaches to the antenna design and full use of the physically available volume and/or full ka sphere [10, 11]. Optimum bandwidth is usually achieved when the radiating electric currents are such that they flow along the outside volume of the sphere defined by the value of ka.

Considering again the short monopole that has increasing capacitive reactance and decreasing radiation resistance, the first objective is to make it resonant (tuning) and the second objective is to then match it to the desired characteristic impedance, typically 50Ω. Tuning of the resonant frequency is easily accomplished by making the short monopole an inverted-L as described in the next section.

1.6.2   The Inverted-L Antenna

A short, straight-wire monopole as described in the previous section has a high capacitive reactance and a low radiation resistance as defined by (1.19). Tuning of the capacitive reactance such that the total feed point reactance X_A is equal to 0 is easily accomplished by increasing the conductor length while at the same time maintaining the same height, h, where $h \ll \lambda$. One of the simplest techniques for tuning a short wire is the inverted-L configuration, as illustrated in Figure 1.13. An alternate view of this configuration is that the straight-wire monopole of height $h = \lambda/4$ is bent over to reduce its height. This maintains nearly the same res-

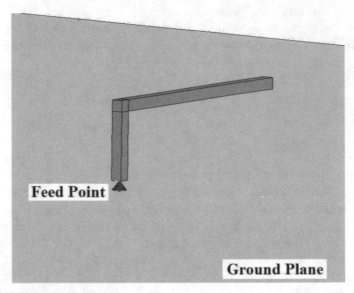

Figure 1.13   Depiction of the inverted-L antenna designed to operate at 2.45 GHz.

onant frequency but the radiation resistance and bandwidth are reduced as a function of the decreasing value of $(h/\lambda)^2$.

If the overall cylindrical diameter of the inverted-L antenna is too large, the horizontal section of wire can simply be meandered in nearly any geometric configuration to reduce the antenna's overall diameter. Meandering does not necessarily have a significant effect on the resonant frequency, which, as necessary, can easily be adjusted by increasing or decreasing the wire length for a decrease or increase in resonant frequency, respectively. The radiation resistance of the meandered inverted-L antenna does not change appreciably, particularly with respect to a 50Ω characteristic impedance, since the value of $(h/\lambda)^2$ does not change appreciably.

A typical curve of impedance versus frequency for an Inverted-L antenna is shown in Figure 1.14, where the low radiation resistance at resonance is evident on the Smith chart. This inverted-L antenna, similar to that depicted in Figure 1.13, has a height of 8 mm and a total conductor length of 29.8 mm. The conductor diameter (width) is 1 mm. It is resonant at approximately 2.45 GHz with a resonant resistance of 7.9 Ω. The inverted-L is mounted over an infinite ground plane. The next design step is to impedance match the inverted-L by making it an inverted-F antenna, as described in the next section.

1.6.3   The Inverted-F and Planar Inverted-F Antennas

Inverted-L antennas as described in the previous section typically have low resonant resistances. A typical curve of impedance versus frequency was presented in Figure 1.14. Any antenna that exhibits this behavior ($R_A(\omega_0) \ll Z_{CH}$) can be impedance matched to Z_{CH} using a parallel inductance connected directly at the feed point. When a relatively large value of parallel inductance is connected at the feed point, there is only a marginal change in the antenna's total feed-point impedance. When a relatively small value of parallel inductance is connected at the feed point, there is a substantial transformation of the antenna's total feed-

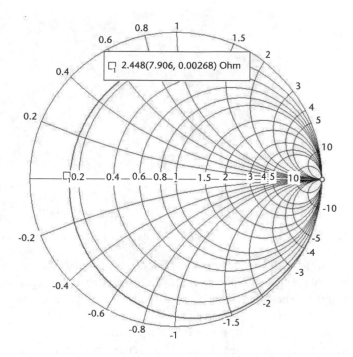

Figure 1.14 Impedance properties of the inverted-L antenna.

point impedance, where it moves towards the short-circuit point on the Smith chart, as illustrated in Figure 1.15.

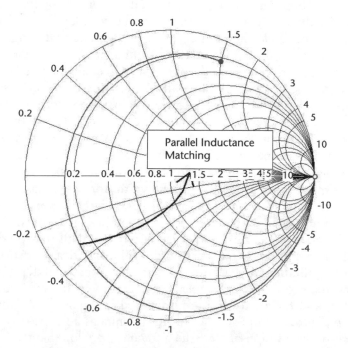

Figure 1.15 Smith chart illustrating the movement of impedance when a parallel inductance is connected at the antenna's feed point. This is the basic behavior associated with impedance matching an inverted-L using a parallel inductor.

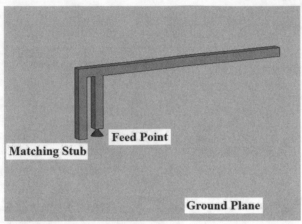

Figure 1.16   Depiction of the inverted-F antenna designed to operate at 2.45 GHz.

If the parallel inductance is added directly at the feed point, the matched frequency will typically be less than the resonant frequency. If the parallel inductance is added to the antenna structure well away from the feed point, the match frequency will typically be greater than the resonant frequency. Tuning or adjustment of the match frequency can be accomplished by movement of the connection point and by adjusting the total wire length within the antenna structure.

A typical implementation of a parallel stub (inductance) used to match an inverted-L antenna is shown in Figure 1.16. This configuration is the familiar inverted-F antenna. Alternate configurations of the parallel inductance impedance matching section (shunt or stub matching) can be easily implemented. For example, the matching stub can be placed in a direction orthogonal to the main horizontal axis of the antenna structure, as will be seen later in a dual-band inverted-F configuration. While these antennas with alternate matching stub configurations may not look like the simple inverted-F antenna, the parallel connection to ground serves the same purpose and the antennas will exhibit similar performance.

The inverted-F antenna illustrated in Figure 1.16 has the same dimensions as the inverted-L of the previous section with the exception of the added parallel matching stub to ground. The antenna is matched at a frequency of approximately 2.45 GHz, with the impedance and VSWR properties, as shown in Figures 1.17 and 1.18, respectively. The VSWR bandwidth (for $s = 3$) of the inverted-F is approximately 8.6%. The value of ka for this antenna is approximately 0.76 and the maximum achievable VSWR bandwidth (for $s = 3$), based on the lower bound on Q, is 31.9%. The antenna's actual bandwidth is approximately one-fourth of the maximum achievable bandwidth. The impedance bandwidths of the inverted-L and inverted-F antennas are substantially less than that predicted by the lower bound since these antenna types do not fill a significant portion of the ka sphere. The 3-D radiation pattern of the antenna is presented in Figure 1.19, which depicts the $|E_{total}|$ over the upper half-space above the infinite ground plane. The $|E_{\theta}|$, $|E_{\phi}|$, and the axial ratio are presented in Figure 1.20. These provide insight into the polarization properties of the antenna as a function of observation angle. At the lower elevation angles, the antenna is predominantly E_{θ} (vertically) polarized. At the upper elevation angles (zenith), the antenna is predominantly E_{ϕ} (horizontally) polarized. In regions between the ground plane horizon and zenith, the

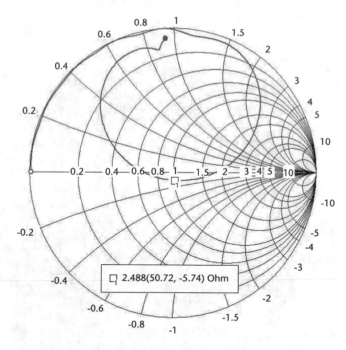

Figure 1.17   Input impedance of the inverted-F antenna.

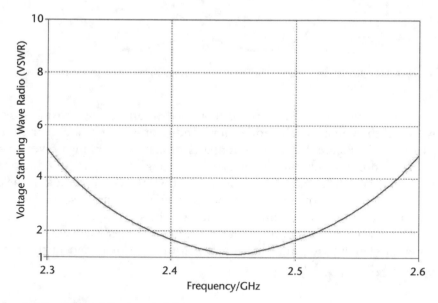

Figure 1.18   VSWR of the inverted-F antenna.

antenna exhibits varying levels of dual polarization. These radiation patterns are presented as reference for understanding changes in the antenna's performance properties as a function of the ground plane size and location of the antenna on the ground plane. These topics will be addressed in the next section.

In many applications, the bandwidth of the inverted-F antenna is not sufficient to cover the desired operating frequency range. In this case, the overall conductor

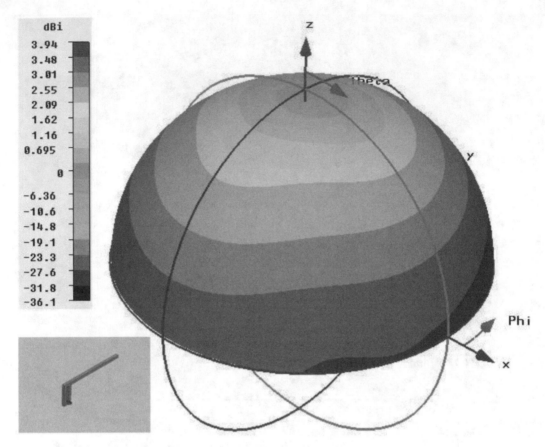

Figure 1.19   Radiation pattern ($|E_{total}|$) of the inverted-F antenna when mounted over an infinite ground plane.

volume of the inverted-F may be increased in an attempt to increase its bandwidth. If the volume of the horizontal conductor is kept to a planar configuration, as shown in Figure 1.21, the antenna is designated a planar inverted-F antenna (PIFA). Adding volume to the inverted-F through the planar conductor at the top of the structure increases the antenna's bandwidth. While adding to the conductor volume, some adjustment of the conductor length in one or several of the horizontal dimensions may be required to maintain the same operating frequency.

Details on the design approaches and techniques used with these and other antenna elements for multiband applications are provided in the remainder of the book.

1.7   Finite Ground Plane Effects on Small Antenna Performance

In nearly all cases, integrated antennas operate with some form of ground plane structure, where the current distribution on this structure is a significant or dominant factor in determining the impedance, bandwidth, radiation patterns, and polarization properties (the operating mode) of the antenna. In the design of integrated antennas, the ground plane structure must often be considered part of the antenna [12].

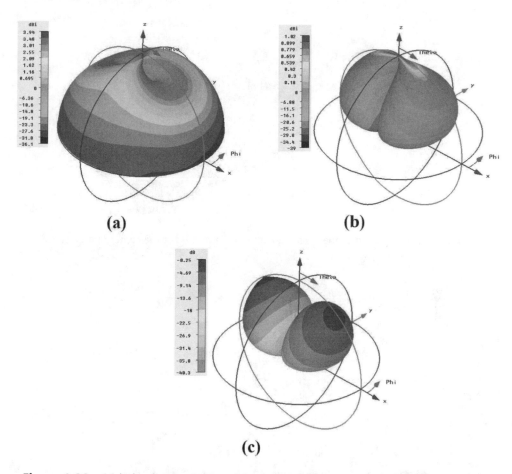

Figure 1.20 (a) $|E_\theta|$ radiation pattern of the inverted-F antenna, (b) $|E_\phi|$ radiation pattern of the inverted-F antenna, and (c) axial ratio of the inverted-F antenna.

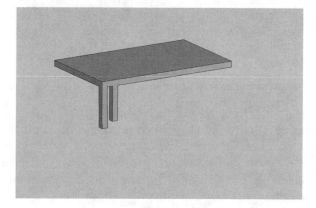

Figure 1.21 The planar inverted-F antenna (PIFA).

Consider the inverted-F antenna described in the previous section. Its imped-ance and radiation pattern properties were considered where the antenna was mounted over an infinite ground plane. In practice, a reasonably sized ground plane may be used, where the impedance properties are usually very similar to

Figure 1.22 Depiction of the inverted-F antenna mounted at the center of a device-sized ground plane.

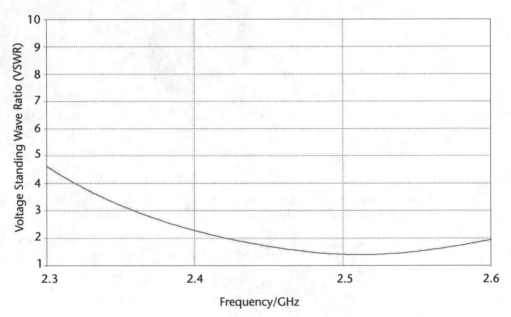

Figure 1.23 VSWR of the inverted-F antenna mounted at the center of the device-sized ground plane.

those where the antenna is mounted over the infinite ground plane but the pattern properties will differ substantially. For example, in the case of the straight-wire monopole on a small ground plane, it is well known that the peak of the radiation pattern will lift off of the ground plane horizon. An omnidirectional radiation pattern is maintained so long as the ground plane is circular or very large.

Figure 1.22 depicts the inverted-F antenna mounted at the center of a device-sized ground plane having dimensions of 100 by 40 mm. The inverted-F's VSWR and radiation pattern ($|E_{total}|$) are presented in Figures 1.23 and 1.24, respectively. The VSWR (impedance) properties are remarkably similar, except for a slight shift the operating band, which is easily compensated for by adjusting the conductor length. While the impedance properties in this case are not so affected by the

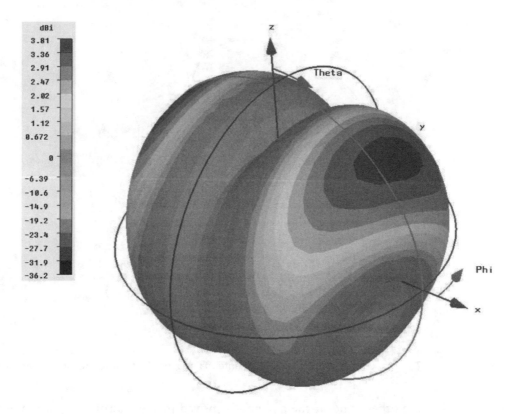

Figure 1.24   Radiation pattern of the inverted-F antenna mounted at the center of the device-sized ground plane.

reduction in the size of the ground plane, there is a substantial change in the radiation pattern. The peak of the main beam is lifted well off the ground plane horizon and the omnidirectional nature of the pattern is substantially diminished for two reasons: one is the obvious asymmetry in the ground plane and the other is the asymmetry in the antenna structure about the feed point, which is not as significant if the ground plane is very large. An important point to note about the pattern is the presence of the overhead (zenith) null. This illustrates that the antenna is operating in a fundamental monopole-mode, where the ground plane size reduction has significantly affected the pattern shape but not the antenna impedance.

Figure 1.25   Depiction of the inverted-F antenna mounted at the edge of the device-sized ground plane.

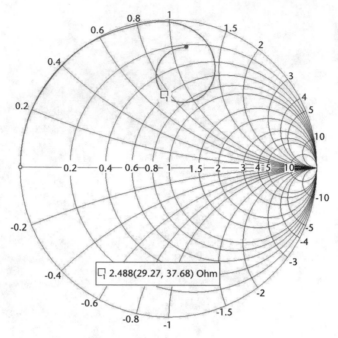

Figure 1.26   Impedance of the inverted-F antenna mounted at the edge of the device-sized ground plane.

On the other hand, if the antenna is moved to the end or edge of the ground plane, as illustrated in Figure 1.25, the basic operating mode of the antenna will change because the currents on the ground plane become more significant in terms of how they affect both the antenna's radiation pattern and impedance properties. These effects can vary substantially as a function of antenna design and operating frequency, where the electrical length of the ground plane is more significant.

The impedance and radiation pattern of the edge mounted inverted-F are presented in Figures 1.26 and 1.27, respectively. Moving the antenna to the edge of the ground plane significantly changes its impedance, bandwidth, and radiation pattern. Most significant in the pattern characteristic is the absence of the overhead (zenith) null. With device-sized ground planes, the antenna's location on the ground plane typically has more of an effect on the antenna's performance than does the size of the ground plane, in and of itself.

A better illustration of these effects is evident when considering an antenna that is designed to operate at a lower frequency. For example, consider the edge-mounted, dual-band inverted-F depicted in Figure 1.28 (note that the parallel matching sections are not in line with the antenna axis but serve the same impedance matching function). There is no optimization here for a specific communications band other than the fact that the antenna was designed to operate with one band at a frequency less than 1 GHz and another band near 2 GHz. While it appears that there are two separate, or perhaps independent radiating structures within the antenna, both are needed to create the two operating bands (if one structure is removed both of the operating bands cease to exist). Here, we only consider the performance properties at the lowest operating band. The return loss properties of the antenna are shown in Figure 1.29.

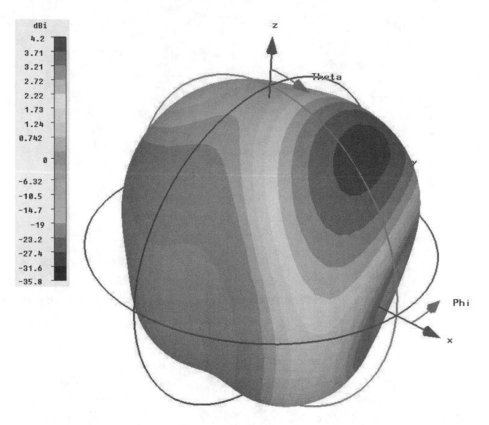

Figure 1.27   Radiation pattern of the inverted-F antenna mounted at the edge of the device-sized ground plane.

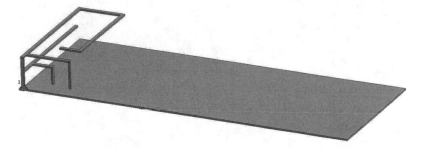

Figure 1.28   Depiction of a dual-band inverted-F antenna mounted at the edge of a device-sized ground plane. The antenna is designed with one operating frequency below 1 GHz and another operating frequency near 2 GHz.

If the antenna is mounted at the center of the device-sized or infinite ground plane, the operating bands cease to exist in both cases, as illustrated in Figures 1.30(a) and 1.30(b), respectively. The radiation patterns of the antenna, when mounted at the center of the device and infinite ground plane, are shown in Figures 1.31(a) and 1.31(b), respectively, for a frequency of 850 MHz. The substantial difference in the radiation pattern is attributed to the size of the ground plane, its rectangular shape, and the asymmetry of the antenna about the feed point,

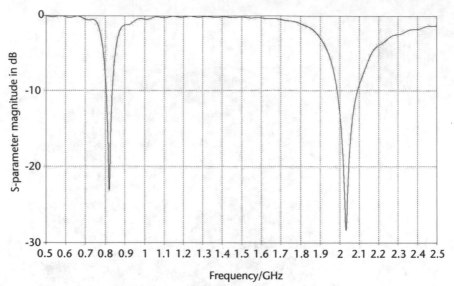

Figure 1.29 Return loss of the dual-band inverted-F antenna mounted at the edge of the device-sized ground plane.

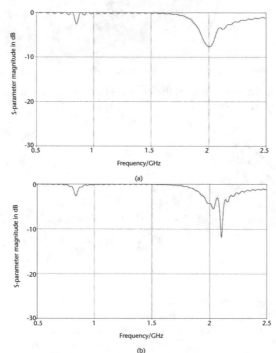

Figure 1.30 Return loss properties of the inverted-F antenna mounted at (a) the center of the device-sized ground plane and (b) the center of an infinite ground plane.

which, as noted earlier, becomes a substantial issue with small ground planes. In both cases, the antenna is fundamentally operating in a monopole mode, with some overhead null fill that arises from the antenna's E_ϕ radiation. In some instances, the level of overhead null fill can be adjusted (reduced) as function of the antenna length, height, and operating frequency.

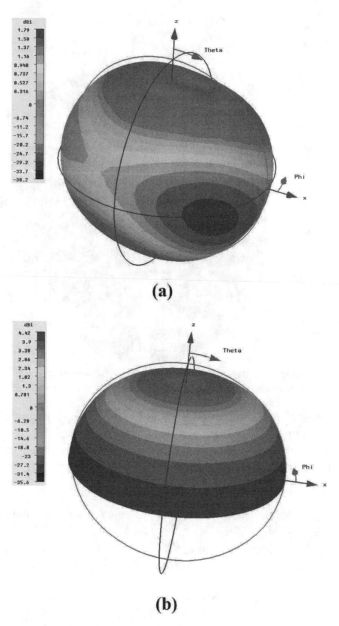

(a)

(b)

Figure 1.31   Radiation pattern of the inverted-F antenna mounted at (a) the center of the device-sized ground plane and (b) the center of an infinite ground plane. The radiation patterns are presented for 850 MHz.

When the antenna is mounted at the edge of the device-sized ground plane, there is a substantial change in the radiation pattern and mode, as illustrated in Figure 1.32. The radiation pattern has well developed nulls aligned with the long axis of the ground plane and it is nearly omnidirectional about the long axis of the ground plane. In this case, the "antenna," which is both the inverted-F and ground plane, operates in a dipole mode, where the radiating dipole current is predominantly on the ground plane. This change in operating mode can be typical in many device-integrated antennas, particularly at the lower frequencies.

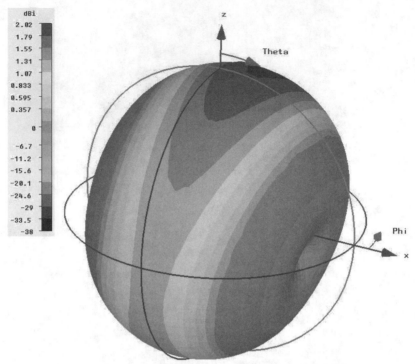

Figure 1.32   Radiation pattern of the dual-band inverted-F antenna mounted at the edge of the
device-sized ground plane. The radiation pattern is presented for 850 MHz.

This general behavior and specifically the change in impedance, also occurs
when an antenna is mounted at the corner of a much larger ground plane. As a
general rule, when edge or corner mounting an antenna, there is an increase in
radiation resistance and bandwidth.

References

[1]   Balanis, C.A., *Antenna Theory: Analysis and Design*, 3rd Edition, John Wiley and
       Sons, Inc., 2005.

[2]   Yaghjian, A. D., and S. R. Best, "Impedance, Bandwidth and Q of Antennas," *IEEE
       Transactions on Antennas and Propagation*, Vol. 53, No. 4, April 2005, pp. 1298–
       1324.

[3]   Chu, L. J., "Physical Limitations of Omni-Directional Antennas," *Journal of Applied
       Physics*, Vol. 10, December 1948, pp. 1163–1175.

[4]   McLean, J. S., "A Re-Examination of the Fundamental Limits on the Radiation Q of
       Electrically Small Antennas," *IEEE Transactions on Antennas and Propagation*, Vol.
       44, May 1996, pp. 672–676.

[5]   Hansen, R. C., "Fano Limits on Matching Bandwidth," *IEEE Antennas and Propaga-
       tion Magazine*, Vol. 47, No. 3, June 2005, pp. 89–90.

[6]   Lopez, A. R., "Review of Narrowband Impedance-Matching Limitations," *IEEE
       Antennas and Propagation Magazine*, Vol. 46, No. 4, August 2004, pp. 88–90,

[7]   Best, S. R., "The Inverse Relationship between Quality Factor and Bandwidth in Mul-
       tiple Resonant Antennas," *Proc. IEEE Antennas and Propagation Society Int. Sympo-
       sium*, July 2006, pp. 623–626.

[8]   Stuart, H. R., S. R. Best, and A. D. Yaghjian, "Limitations in Relating Quality Factor to Bandwidth in a Double Resonance Small Antenna," *IEEE Antennas and Wireless Propagation Letters*, Vol. 6, 2007, pp. 460–463.

[9]   Best, S. R., "A Discussion on the Performance Properties of Electrically Small Self-Resonant Wire Antennas," *IEEE Antennas and Propagation*, Vol. 46, No. 6, December 2004, pp. 9–22.

[10]   Best, S. R., "The Radiation Properties of Electrically Small Folded Spherical Helix Antennas," *IEEE Transactions on Antennas and Propagation*, Vol. 52, No. 4, April 2004, pp. 953–960.

[11]   Best, S. R., "Low Q Electrically Small Linear and Elliptical Polarized Spherical Dipole Antennas," *IEEE Transactions on Antennas Propagation*, Vol. 53, No. 3, March 2005, pp. 1047–1053.

[12]   Best, S. R., "A Discussion on Small Antennas Operating with Small Finite Ground Planes," *Proc. IEEE International Workshop on Antenna Technology: Small Antennas and Novel Metamaterials*, March 2006, pp. 152–155.

CHAPTER 2

Multiband Multisystem Antennas in Handsets

Kevin Boyle

2.1 Introduction

This chapter is concerned with mobile antennas that are designed to operate in a number of frequency bands. Several different systems may also be operational at any one time, usually, but not necessarily, in different bands. In this introductory section we will first outline some of the most common frequency bands and systems. Typical performance specifications for multiband, multisystem antennas are also given.

Section 2.2 deals with some basic techniques—commonly used within the mobile phone industry—of achieving multiband operation. Section 2.3 considers some techniques that can be used to avoid intersystem interference.

2.1.1 Frequencies and Systems

Prior to the 1990s, first generation (1G) analog mobile phones operated in a single frequency band. In Europe, for example, the Total Access Communication System (TACS) operated in a band from 880 to 960 MHz, whereas in North America the Advanced Mobile Phone System (AMPS) operated in a band from 824 to 894 MHz. Hence, early antenna designs were single-band and relatively narrowband. We define the fractional bandwidth of a system as

$$B_F = \frac{f_2 - f_1}{\sqrt{f_1 f_2}} \times 100 \qquad (2.1)$$

where f_1 and f_2 are the lower and upper frequencies of the band, respectively. Using this definition, the fractional bandwidths of AMPS and TACS are approximately 9%. Both use a frequency division duplex (FDD): for TACS 880 to 915 MHz is used for transmission and 925 to 960 MHz is used for reception, whereas for AMPS, 824 to 849 MHz and 869 to 894 MHz are used for transmission and reception, respectively.

First generation systems were largely incompatible: a phone bought in one country could not be used in another. This changed with the introduction of digital second generation (2G) systems in the 1990s. The Global System for Mobile Communications (GSM) was introduced in Europe and subsequently much of the world and a digital version of AMPS—the Interim Standard 54 (IS-54)—was introduced in America. The latter evolved into IS-136 and is often referred to as

D-AMPS (digital AMPS) or TDMA (since it employs a time division multiple access scheme). GSM allowed global operation in the original TACS band from 880 to 960 MHz, and then later in a band from 1,710 to 1,880 MHz, leading to a requirement for dual-band operation. However, these frequency bands were not available worldwide: in particular, a different band (1,850–1,980 MHz) was utilized in the United States for IS-54, IS-136, a third system, IS-95, and eventually GSM. This led to the introduction and widespread adoption of tri-band GSM phones. Indeed, quad-band GSM phones eventually emerged, as the frequency bands associated with AMPS in the United States (824–894 MHz) were reallocated to GSM (among other systems).

With the aim of increased mobility and functionality compared to their 2G predecessors, third generation (3G) systems were introduced, first in Japan in 2001, then in Europe in 2003, and subsequently in many other countries worldwide. When second generation systems were introduced, they largely replaced first generation systems. However, 3G systems were introduced to sit alongside existing 2G systems. New spectrum was allocated for 3G services in Europe, Japan, and many other countries, offering the prospect of increased total capacity. However, once again common new spectrum was not available worldwide, which led to a number of frequencies being allocated for 3G, many of which coincide with the frequencies that are used for 2G. Where this happens, it is expected that the 2G systems will eventually be phased out. The International Telecommunication Union (ITU)—the international body that regulates radio and telecommunications worldwide (and an agency of the United Nations)—allowed several different types of systems to be called 3G under the umbrella of International Mobile Telecommunications-2000 (IMT-2000). Two worldwide variants of 3G emerged, one based on code division multiple access (CDMA) and another based on wideband code division multiple access (WCDMA). The latter is based on work that began in the early 1990s towards a successor of GSM named the Universal Mobile Telecommunications System (UMTS). This later became known as UTRA (UMTS Terrestrial Radio Access)—a term used to describe the UMTS radio access solution. The UTRA 3G frequencies are specified in Table 2.1. Shared 2G/3G bands are indicated.

Common combinations of bands are as follows:

- Tri-band GSM: In bands VIII, III, and II
- Quad-band GSM: In bands V, VIII, III, and II
- Triband GSM, single-band UMTS: GSM in bands VIII, III, and II. UMTS in band I
- Quad-band GSM, triband UMTS: GSM in bands VIII, III, II, and I. UMTS in band V, II, and I.

The bands involved are shown in Figure 2.1.

At around the same time that 3G was launched on the market, Bluetooth radios were introduced into phones to allow short-range communications with other devices in an industrial, scientific and medical (ISM) band centered at 2.4 GHz. In effect, this meant that some phones became tri-mode. Since the different modes may also operate in different frequency bands, phones also became increasingly multi-band.

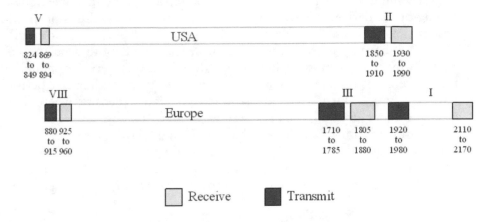

Figure 2.1 Common cellular frequency bands (MHz) used in Europe and the United States, with their UTRA band designations.

Table 2.1 UTRA Frequency Bands

Band	TX (MHz)	RX (MHz)	Notes
I	1,920–1,980	2,110–2,170	UMTS in Europe and Asia
II	1,850–1,910	1,930–1,990	Used for 2G, particularly in the United States
III	1,710–1,785	1,805–1,880	Widely used for 2G, particularly GSM
IV	1,710–1,755	2,110–2,155	Allocated for 3G in the United States
V	824–849	869–894	Used for 2G, particularly in the United States
VI	830–840	875–885	Allocated for 3G in Japan
VII	2,500–2,570	2,620–2,690	Worldwide UMTS extension band
VIII	880–915	925–960	Widely used for 2G, particularly GSM
IX	1,749.9–1,784.9	1,844.9–1,879.9	Used in Japan
X	1,710–1,770	2,110–2,170	

This trend towards the inclusion of more systems operating in more bands continued with the introduction of the Global Positioning System (GPS), Wi-Fi and new systems such as mobile television (with several standards and bands proposed) and WiMAX (again with several bands). Importantly, many of these systems may be required to operate simultaneously.

Multimode, multiband operation presents a formidable challenge to mobile phone designers, particularly for the RF parts. Of these, the antennas occupy the largest volume and, hence, have the biggest impact on the (commercially crucial) styling of the device. It is well-known that antenna bandwidth is proportional to volume, as discussed in Section 1.4. However, the space allocated to the antenna(s) has not increased: on the contrary, it is tending to decrease due to the demand for thinner, more highly stylized devices. If this were not enough, there is also increasing pressure on mobile phone manufacturers to improve performance, particularly in countries with areas of low population density, where, due to the large distances involved, cellular coverage is often "patchy."

It can be seen from Figure 2.1 that quad-band GSM coverage can be achieved using a dual-band GSM900/GSM1800 design with extended bandwidth in both bands. An approximate doubling of the bandwidth is required. By further extending the high-band, it is also possible to cover six UTRA bands.

2.1.2 Typical Specifications

Mobile phone antennas are typically designed to have a return loss of better than 6 dB. This corresponds to a VSWR of 3:1 and a reflected power of 25% (or an equivalent transmission—which can be thought of as an efficiency due to mismatch—of 75%). This is a relatively high level of reflection (or low level of transmission) when compared to other applications, which, to some extent, reflects the difficulty of achieving a good match over a number of bands with an electrically small antenna. Often the phone aesthetic design will have priority over electrical considerations: for example, it would be unthinkable for an antenna designer to argue for a slightly larger antenna should thin phones be demanded by the end consumer (or by the marketeers' perception of what would inspire consumers to buy in large numbers and at good profit margins). Normally the average return loss is better than 6 dB since very often only the "edges" of the frequency band (f_1 and f_2 as defined above) give the worst case.

The antenna radiation efficiency is generally required to be greater than 50%. Sometimes this includes the loss due to mismatch and sometimes this is only an average figure (averaged over each frequency band), dependent on the particular requirements of the handset manufacturer. Sometimes these requirements may change, dependent on the usage of the phone and the mechanical design: for example, when a particularly small phone design is specified, this may be accompanied by a relaxation of the overall efficiency requirement. This is often necessary, since the antenna bandwidth is a strong function of the PCB ground plane size, as discussed in Section 1.7.

2.2 Multiband Techniques

The requirement for multiband operation began with the widespread use of GSM in two frequency bands. In Europe, two widely separated bands—band III and band VIII—were the first to be used. This introduced a need to design an antenna with a single feed that could operate in two relatively narrow bands, with one band centered at a frequency that is approximately twice that of the other. We will show how this can be achieved in the following sections.

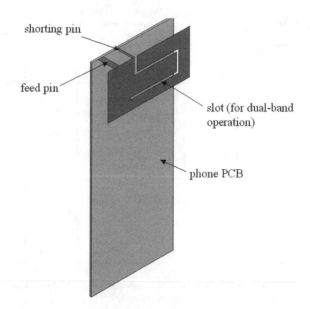

Figure 2.2   Dual-band PIFA geometry.

A dual-band planar inverted F antenna (PIFA) and a supporting PCB is shown in Figure 2.2. The PIFA is probably the most commonly used mobile phone antenna due to its reasonable use of space—RF components may be mounted beneath it—and low specific absorption rate (SAR). It is derived from the "shunt-driven inverted L antenna" described by King et al. in 1960 [1]. It is often argued that the inner part of this structure provides a high frequency resonance, whereas the outer part provides a low-frequency resonance. Here, a radiating and balanced mode analysis is applied that allows equivalent circuits representing the radiating and reactive contributions of dual-band PIFAs to be identified. In turn, this gives an improved insight into how this class of antenna works: in particular, it shows that the slot does not separate two resonators, rather that it provides reactive loading. The example antenna of Figure 2.2 is used to illustrate a simple design procedure based on the proper selection of slot and shorting pin dimensions in Sections 2.2.2 and 2.2.3. However, we must first introduce the basic theory.

2.2.1   Radiating and Balanced Mode Theory

Radiating and balanced mode theory was developed by Uda and Mushiake [2] to analyze closely coupled antenna structures and has subsequently been widely used to analyze the folded monopole and dipole. The theory presented below is similar to that used for a folded monopole, but adapted to take account of the effect of the PCB and the planar section of the PIFA.

Here the analysis is generalized to include an arbitrary load impedance. The special cases of short-circuit and open-circuit loads are used to analyze the effects of the shorting pin and slots in the planar section of the antenna, respectively. An equivalent circuit, that includes the effect of the PCB, is presented.

A PIFA and an associated PCB/handset are shown diagrammatically in Figure 2.3 [3,4]. The PIFA is shown on the top of the PCB/handset for clarity, but would

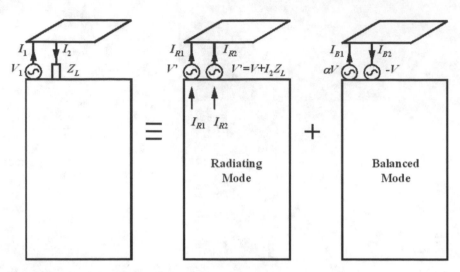

Figure 2.3　Radiating and balanced mode analysis of a PIFA with a loaded unfed pin. (From: [4]. ©2006 IEEE. Reprinted with permission.)

normally be in the same plane as the PCB—this does not affect the following theory (which can also be applied to monopoles of the type used in some so-called clamshell phones). The PCB is assumed to be a good conductor: this is reasonable in practice due to the presence of a substantial ground plane and the parasitic coupling between densely spaced interconnects.

When the feed and shorting conductors are electrically close, this configuration can be decomposed into radiating and balanced modes as shown. Here the problem is generalized to allow the pin that is normally shorted to be arbitrarily loaded by a complex impedance, Z_L. This is incorporated in the radiating and balanced mode analysis by replacing it with a voltage source of the same magnitude and polarity as the voltage drop across the load.

To prevent the balanced mode from contributing to radiation it is necessary to pay attention to the current sharing between the feed and shorting pins. The balanced mode excitation voltages should be such that equal and opposite currents are maintained (i.e., $I_{B1} = I_{B2} = I_B$). The reader is referred to [5] for a method of achieving this in a computationally efficient manner, without any prior knowledge of the geometry of the structure. Since the source voltages are the same in the radiating mode and effectively connected in series in the balanced mode, Figure 2.3 may be simplified, as shown in Figure 2.4. This simplified equivalent makes it clearer how radiating and balanced mode impedances, Z_R and Z_B, respectively, are derived.

In the radiating mode it is assumed that, due to strong mutual coupling, the currents in the feed and shorting pins differ by a current sharing factor, α given by

$$\alpha = \frac{I_{R2}}{I_{R1}} \tag{2.2}$$

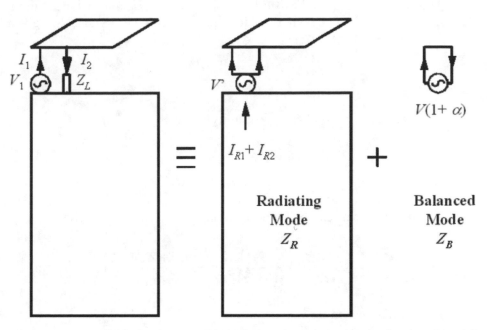

Figure 2.4   Simplified radiating and balanced mode analysis of a PIFA showing Z_R and Z_B.

This depends primarily on differences in the cross sections of the fed and loaded pins. If the pins differ, the one with the larger cross section will carry the most current. The input current, I_1 is given by

$$I_1 = I_{R1} + I_B = \frac{V'}{(1+\alpha)Z_R} + \frac{V(1+\alpha)}{Z_B}$$ (2.3)

where
    V balanced mode voltage
    V' radiating mode voltage
    Z_R impedance of the dual feed pin PILA and PCB combination, derived from the radiating mode
    Z_B impedance of the short circuit transmission line formed between the feed and shorting pins, derived from the balanced mode
The radiating mode voltage is given by

$$V' = V + I_2 Z_L = V + (I_B - \alpha I_{R1})Z_L$$ (2.4)

Using the two terms in (2.3), grouping terms in V and V' and simplifying gives

$$V' = V\frac{(1+\alpha)Z_R Z_B + (1+\alpha)^2 Z_R Z_L}{(1+\alpha)Z_R Z_B + \alpha Z_L Z_B}$$ (2.5)

Thus, a relation is established between the radiating and balanced mode voltages. A relation can also be derived for the input voltage, V_1, which is given by

$$V_1 = V' + \alpha V \tag{2.6}$$

Substituting (2.5) in (2.6) and simplifying gives

$$V_1 = V \frac{(1+\alpha)^2 Z_R (Z_L + Z_B) + \alpha^2 Z_L Z_B}{(1+\alpha) Z_R Z_B + \alpha Z_L Z_B} \tag{2.7}$$

The input current can be found from (2.3) and (2.5) and is given by

$$I_1 = V \frac{(1+\alpha)^2 (Z_L + Z_R) + Z_B}{(1+\alpha) Z_R Z_B + \alpha Z_L Z_B} \tag{2.8}$$

Hence, the impedance is given by

$$Z_1 = \frac{(1+\alpha)^2 Z_R (Z_L + Z_B) + \alpha^2 Z_L Z_B}{(1+\alpha)^2 (Z_L + Z_R) + Z_B} \tag{2.9}$$

For the special case of $Z_L = 0$, this simplifies to

$$Z_1 = \frac{(1+\alpha)^2 Z_R Z_B}{(1+\alpha)^2 Z_R + Z_B} = \frac{K_{\alpha s} Z_R Z_B}{K_{\alpha s} Z_R + Z_B} \tag{2.10}$$

This is similar to the well-known expression for a folded dipole—the radiating mode is impedance transformed by a factor $K_{\alpha s}$ and adds in parallel with the balanced mode.

Also of interest is the open circuit case. Setting $Z_L = \infty$ gives

$$Z_1 = Z_R + \left(\frac{\alpha}{1+\alpha}\right)^2 Z_B = Z_R + K_{\alpha o} Z_B \tag{2.11}$$

In this case, the effect on the impedance is almost opposite to that when a short circuit is applied. The radiating mode impedance is not transformed, while the balanced mode impedance is transformed by a factor $K_{\alpha o}$. Most importantly, the radiating and balanced mode impedances now add in series rather than in parallel, as with a short circuit.

The results obtained for short-circuit and open-circuit loads are fundamental to the operation of single band and multiband PIFAs. The short circuit case can be used to analyze the effect of the shorting pin, while the open-circuit case can be used to show the influence of slots in the top plate. The detail of this behavior is

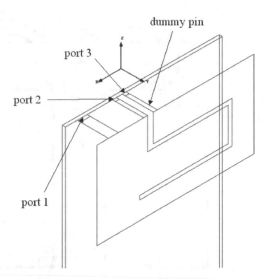

Figure 2.5   Dual-band PIFA showing with dummy pin for analysis of the slot. (From: [6]. ©2005 IEEE. Reprinted with permission.)

shown in the following sections. Slots and shorting pins are dealt with in Sections 2.2.2 and 2.2.3, respectively.

2.2.2 Analysis of PIFA Slots

Here the effect of the slot is treated using a radiating and balanced mode analysis with the open-circuit loading condition. Consider the geometry shown in Figure 2.5.

This is the same as the antenna shown in Figure 2.2, except here the antenna has three feeds (ports). Feed 3 and its associated pin are "dummy" elements, for the purposes of studying the effect of the slot. In the final design they are removed. Feed 1 is connected to the radio and feed 2 is shorted in the final design. A radiating and balanced mode analysis is performed as illustrated in Figure 2.6.

The analysis starts by connecting feeds 1 and 2 together and applying radiating and balanced mode voltages to feeds 1 and 2 (together) and feed 3. Then (2.11) is used to simulate the condition where feed 3 is open circuit via the summation of radiating and balanced modes. The impedances associated in the radiating and transformed balanced modes are given in Figure 2.7.

In both bands the radiating mode impedance, Z_R equivalent to that of a planar-inverted L antenna (PILA) without a slot or shorting pin, indicating that the slot has little effect on the radiating mode at these frequencies.

In the balanced mode the slot acts as a short circuit transmission line that is below quarter-wave, and hence inductive, in band VIII and above quarter-wave, and hence capacitive, in band III. Since the natural resonant frequency of the antenna (in the radiating mode or as a PILA) is somewhere between the two bands, this allows the antenna to achieve dual-band operation with proper choice of the slot length and current sharing factor (the radiating and balanced modes add in series with an open circuit load). The current-sharing factor is determined by the relative widths of the conductors on either side of the slot. It can be seen from Figure 2.7 that the slot length and current sharing factor have been opti-

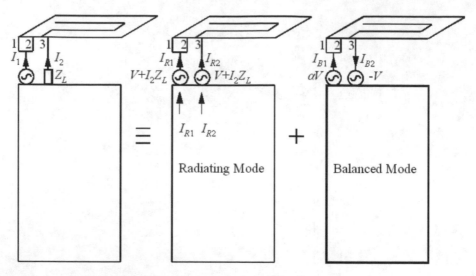

Figure 2.6   Radiating and balanced mode analysis of the slot.

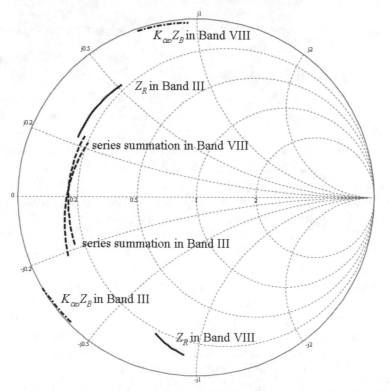

Figure 2.7 S_{11} of the PIFA configuration of Figure 2.5 in the open-circuit radiating, balanced, and sum modes (with an open circuit simulated at port 3).

mized such that the summation of the radiating and balanced modes gives resonance in bands III and VIII.

This analysis is contrary to the commonly held view that the slot facilitates a
separation between two resonant lengths in the top plate. It can be used to explain

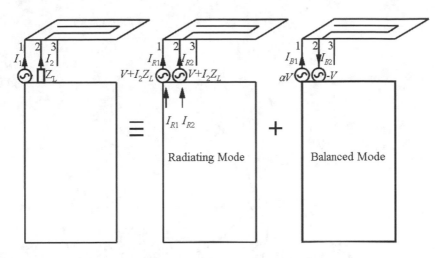

Figure 2.8   Radiating and balanced mode analysis of the shorting pin.

why there is a trade-off between the bandwidths obtained in the two bands, as previously observed, for example, in [7]. It is important to note that the whole structure is used at both frequencies in the radiating mode. The slot simply provides tuning reactance.

2.2.3   Analysis of PIFA Shorting Pins

The effect of the shorting pin can be analyzed by applying radiating and balanced voltages to feed 2 and feed 1 while feed 3 is open circuit, as shown in Figure 2.8.

Here feed 2 is the source and feed 1 is shorted. The resulting radiating and balanced mode impedances, derived using the short-circuit load condition given in (2.10), are shown in Figure 2.9. The radiating mode is shown with and without the impedance transformation (determined by the factor $K_{\alpha s}$).

The radiating mode is that of a slotted PILA. The balanced mode is that of a short-circuit transmission line. As seen previously, the radiating and balanced modes add in parallel. A simplified wideband circuit representation of the dual-band PIFA and PCB combination is shown in Figure 2.10.

Here the series components L_A, C_A and R_A represent the fundamentally series resonant impedance of the antenna (without any slots or shorting pins and at frequencies substantially below the first antiresonance). The contribution of the PCB is modeled as a series resonant circuit (L_{P3} and C_{P3}) connected in series with two parallel resonant circuits. These parallel circuits are used to represent the half-wave and full-wave antiresonances of the PCB. Higher-order modes are not usually important for mobile radio frequency bands and are, therefore not modeled. The series connected parallel circuit formed by L_S and C_S represents the effect of the slot in the PILA top plate. Finally, the transformer and shunt inductance, L_P, represent the effect of the shorting pin (in the radiating and balanced modes, respectively), which, importantly, is effective after the slot. It should be clear that the impedance transformation acts on the impedance contributions of both the (slotted) antenna and the PCB.

Over a narrow band the equivalent circuit of Figure 2.10 can be simplified by modeling both the antenna (without any slots or shorting pins) and the PCB as a single series resonant circuit. For the 100×40×1-mm PCB considered above, the

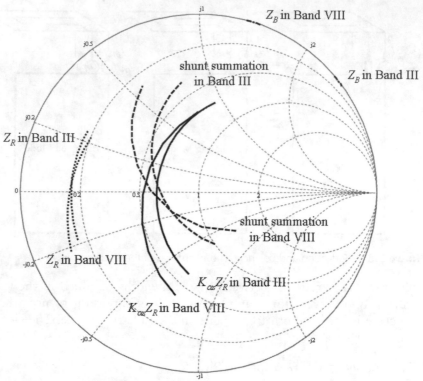

Figure 2.9 S_{11} the PIFA configuration of Figure 2.5 shown in the short-circuit radiating, balanced, and sum modes (with a short simulated at port 2 and an open circuit at port 3).

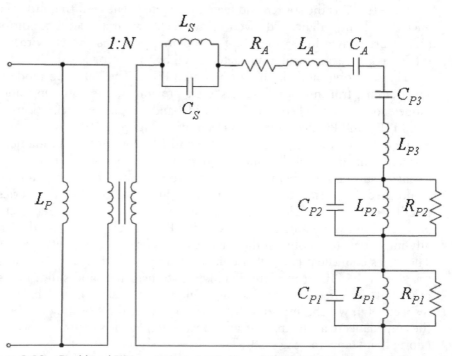

Figure 2.10   Dual-band PIFA equivalent circuit. (From [4]. ©2006 IEEE. Reprinted with permission.)

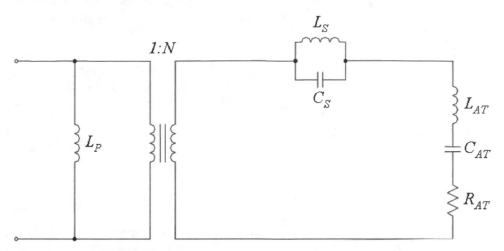

Figure 2.11   Simplified dual-band PIFA equivalent circuit. (From [4]. ©2006 IEEE. Reprinted with permission.)

resistance of this circuit is approximately the same in bands III and VIII. Hence a simplified equivalent circuit can be used to represent both bands, as shown in Figure 2.11.

For such a circuit, it can be shown that equal bandwidth can be achieved in the GSM and DCS bands when the series components (L_{AT}, C_{AT}, and R_{AT}), representing the basic PILA, and the parallel components (L_S and C_S), representing the slot, are resonant at the geometric mean of the band III and VIII center frequencies. If the resonant frequency of the PILA is lower than the geometric mean, the band VIII bandwidth is improved at the expense of that at band III. The converse applies for higher frequencies. Hence, the bandwidths of the two bands are not independent; to improve one band the other band must be degraded. The basic dimensions of the antenna, the positions of the feed and shorting pins and the length and position of the slot can be chosen according to the bandwidth requirement.

2.3   Multisystem Techniques

Consider the systems shown in Figure 2.12. There are two antennas and two sources (or receivers), both of which can be active at the same time. By superposition, the systems can be decomposed as shown. This allows us to more easily consider the operation of each system in the presence of the other.

In most situations encountered in practice, each system operates at a different frequency, allowing them to be filtered against each other, as shown in Figure 2.12. In the following commentary we use GSM and Digital Video Broadcasting—Handheld, (DVB-H) as examples of systems that could operate simultaneously at different frequencies. The example is based on European frequency allocations: GSM is used in the 880 to 960 MHz band and DVB-H is likely to be used in the 470 to 700 MHz band. DVB-H is a terrestrial digital standard for bringing broadcast services such as TV to low-power, portable receivers. At the portable device, it is used only for reception.

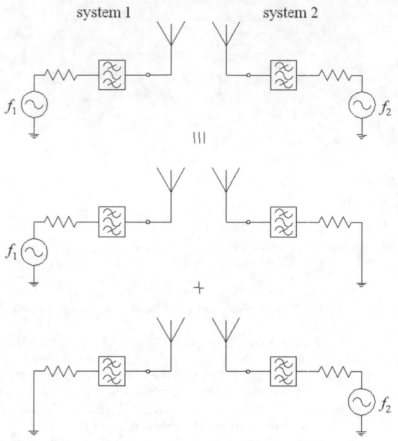

Figure 2.12 Simultaneously operated antenna systems.

If a GSM antenna is transmitting, some of the power will be received by the DVB-H antenna and the circuitry to which it is attached. We require this power to be low to maintain a high GSM efficiency and to prevent the DVB-H receiver from becoming saturated. In other words, we require the isolation between the two systems to be high. The isolation is given by

$$Isolation = \frac{P_T}{P_R} \qquad (2.12)$$

where P_T is the power transmitted by the system under consideration and P_R is the power received by the other system. The isolation is minimized when both systems are conjugately matched. Under such conditions the isolation is determined only by the antennas (i.e., by their relative geometries). For mobile antennas this isolation is often low because part of each antenna—the phone PCB—is shared. The total isolation will also depend on reflections at the antenna to circuit interfaces, as shown in Figure 2.13. The transmit antenna will normally have close to a conjugate match for maximum efficiency; hence, we shall assume that ρ_1 is zero.

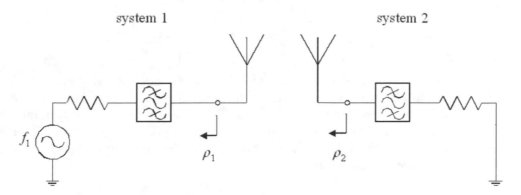

Figure 2.13 Antenna and receiver equivalent simplified circuit.

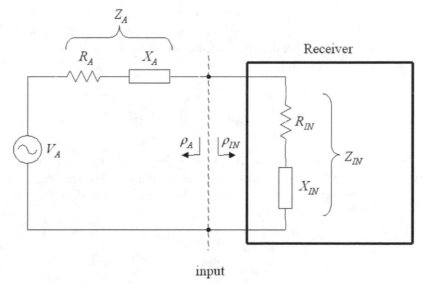

Figure 2.14 Antenna and receiver equivalent simplified circuit.

To analyze the power reflection that occurs at the receive antenna to receiver interface (system 2 in Figure 2.13), and hence the effect on isolation, consider Figure 2.14.

Figure 2.14 shows an equivalent circuit of the DVB-H antenna and receiver (system 2 in Figure 2.13) at the GSM frequency. The antenna is out-of-band, and will therefore have an impedance, Z_A, that is likely to be mismatched. Similarly, the receiver will be out-of-band and will also be mismatched. The band filter within the receiver normally determines the degree of mismatch. If we choose an arbitrary system impedance (normally 50 ohms), we can represent the antenna and receiver mismatches by the antenna and receiver reflection coefficients, ρ_A and ρ_{IN}, respectively.

We are interested in the power transferred at the receiver input—or the lack of it—from the GSM antenna. Here the source is an antenna and the load is the receiver, but the analysis could apply to the interface between any two circuits. The transmission coefficient from source to receiver is given by

$$|\tau_2|^2 = \frac{\left(1-|\rho_A|^2\right)\left(1-|\rho_{IN}|^2\right)}{|1-\rho_A\rho_{IN}|^2} \qquad (2.13)$$

For a given $|\rho_A|$ and $|\rho_{IN}|$, the numerator is constant and represents the combined isolation due to mismatch of the antenna and receiver when measured independently with 50Ω terminations. The minimum and maximum of the denominator occur when the phases of ρ_A and ρ_{IN} add to give 0 and 180 degrees, respectively. When the denominator is at its minimum, there is most transmission and, hence, lowest isolation. The condition for this is

$$\arg\{\rho_{IN}\} = -\arg\{\rho_A\} \qquad (2.14)$$

When this occurs, the isolation of the antenna-receiver combination is less than the independently calculated isolation due to mismatch (with 50Ω terminations) by a factor

$$\left(1-|\rho_A||\rho_{IN}|\right)^2 \qquad (2.15)$$

This "loss" of isolation can be significant when the product $|\rho_A||\rho_{IN}|$ is large, as can occur when both the antenna and receiver are operated out-of-band. When the denominator is maximized, the isolation is optimized. The condition for this is that

$$\arg\{\rho_{IN}\} = 180 - \arg\{\rho_A\} \qquad (2.16)$$

Here the isolation of the antenna-receiver combination is greater than the independently calculated isolation due to mismatch by a factor

$$\left(1+|\rho_A||\rho_{IN}|\right)^2 \qquad (2.17)$$

This isolation "gain" cannot be greater than 6 dB, since the product of $|\rho_A|$ and $|\rho_{IN}|$ cannot be greater than unity.

The ratio of the maximum and minimum transmission coefficients is given by

$$\frac{|\tau_2|^2_{max}}{|\tau_2|^2_{min}} = \left(\frac{1+|\rho_A||\rho_{IN}|}{1-|\rho_A||\rho_{IN}|}\right)^2 \qquad (2.18)$$

This shows that when the product $|\rho_A||\rho_{IN}|$ is large, variations in the phase of both $|\rho_A|$ and $|\rho_{IN}|$ can result in big differences in the level of isolation achieved.

If, for example, we take $|\rho_A||\rho_{IN}|$=0.9, there is a 25-dB variation in the level of isolation that can be achieved. With $|\rho_A||\rho_{IN}|$=0.7, the difference is 15 dB. Clearly, this difference is greatest when both the antenna and circuit are reflective. The deviation from the maximum isolation with phase can be found from

$$\Delta\left|\tau_2\right|^2 = \frac{\left|\tau_{IN}\right|^2}{\left|\tau_{IN}\right|^2_{min}} = \left(\frac{1+\left|\rho_A\right|\left|\rho_{IN}\right|}{\left|1-\rho_A\rho_{IN}\right|}\right)^2 \qquad (2.19)$$

With the substitution $\kappa = |\rho_A||\rho_{IN}|$, this can be written

$$\Delta\left|\tau_2\right|^2 = \frac{\kappa^2 + 2\kappa + 1}{\kappa^2 - 2\kappa\cos\psi + 1} \qquad (2.20)$$

where

$$\psi = \arg\{\rho_A\} + \arg\{\rho_{IN}\} \qquad (2.21)$$

This is plotted in Figure 2.15.

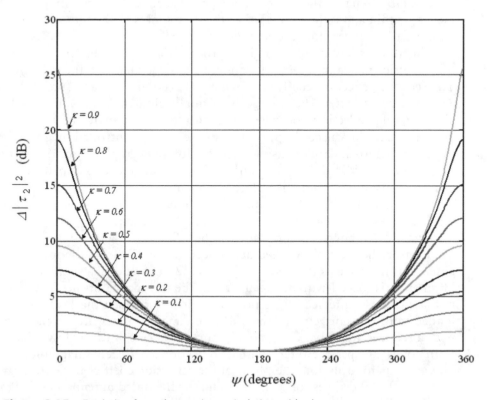

Figure 2.15   Deviation from the maximum isolation with phase.

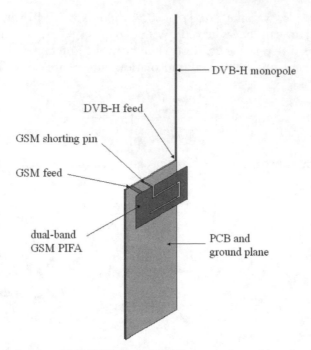

Figure 2.16   A dual-band GSM PIFA and a DVB-H monopole on a handset PCB.

It can be seen that the curve is quite broad around the optimum phase of 180°. Hence, within approximately ±50° of the optimum is always good enough to give within 1 dB of the maximum possible isolation.

The above shows that for a given product of the receiver and antenna reflection coefficient magnitudes, the isolation is dependent on the relative phases. The receiver reflection coefficient magnitude is limited by the filter that is used. The larger the better—how large is normally limited by the filter technology. The antenna reflection coefficient magnitude is limited by the antenna design and the technology used. Importantly, phase can be controlled by good choice of filter characteristics and/or using transmission lines or phase-shifting networks.

Having considered the effect of the receiving antenna load impedance on isolation, we must also consider its effect on the impedance characteristic of the transmitting antenna. We shall do this using a dual-band GSM PIFA and a DVB-H monopole—shown in Figure 2.16—as example antennas.

Figure 2.17 shows the impedance of the GSM antenna in band VIII with either an open or short circuit applied at the feed of the DVB-H monopole. The GSM antenna is the same as that used in Section 2.2. It can be seen that when the DVB-H antenna is loaded with an open circuit, the impedance of the GSM antenna is not significantly affected by its presence. However, when the DVB-H antenna is loaded with an short circuit, the GSM antenna resonates at a higher frequency with a reduced resistance and, more importantly, a significantly lower bandwidth. Both loads are perfect filters with a reflection coefficient magnitude of unity. However, as for isolation, the phase of the reflection coefficient (zero for an open circuit and 180 degrees for a short circuit) at the unfed antenna has a significant effect on the impedance match of the fed antenna.

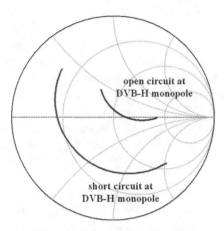

Figure 2.17   Impedance of the GSM PIFA of Figure 2.16 in band VIII with open and short circuit loads applied to the feed of the DVB-H antenna.

To alleviate such problems, filters can be built into the antenna structure. An example, where a GSM blocking filter is built into a DVB-H monopole using the radiating and balanced mode theory developed in Section 2.2.1, is shown in Figure 2.18 [8].

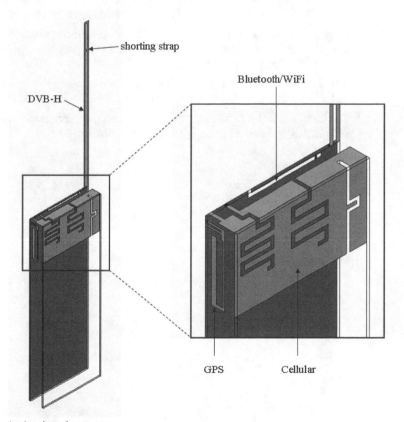

Figure 2.18   A nine-band antenna arrangement.

Here the DVB-H antenna is folded with an open circuit load applied to the loaded arm. Hence, (2.11) applies. The shorting strap on the antenna is chosen such that the balanced mode transmission line is antiresonant at for GSM in band VIII. Hence the DVB-H antenna effectively has an open circuit load at GSM frequencies. This satisfies the loading requirements for both good matching of the GSM antenna and good isolation between the GSM and DVB-H antennas.

The design shown in Figure 2.18 is capable of operating in five cellular bands, two DVB-H bands, the GPS band, and the Bluetooth/Wi-Fi band. Up to four systems can operate simultaneously with only moderately reduced efficiencies.

References

[1]   King, R., C. W. Harrison, and D. H. Denton, "Transmission Line Missile Antennas," *IRE Transactions on Antennas and Propagation*, Vol. 8, January 1960, pp. 88–90.

[2]   Uda, S., and Y. Mushiake, *Yagi-Uda Antennas*, Tokyo: Maruzen Co., 1954.

[3]   Boyle, K. R., "Antennas for Multi-Band RF Front-End Modules," Ph.D. Thesis, Delft University Press, 2004.

[4]   Boyle, K.R., and L.P. Ligthart, "Radiating and Balanced Mode Analysis of PIFA Antennas," *IEEE Transactions on Antennas and Propagation*, Vol. 54, No. 1, January 2006, pp. 231–237.

[5]   Boyle, K. R., "Radiating and Balanced Mode Analysis of PIFA Shorting Pins," *Proc. IEEE Antennas and Propagation Society Int. Symposium and USNC/URSI National Radio Science Meeting*, June 2002, Vol. 4, pp. 508–511.

[6]   Boyle, K. R., and L. P. Ligthart, "Radiating and Balanced Mode Analysis of User Interaction with PIFAs," *Proc. IEEE Antennas and Propagation Society Int. Symposium*, July 2005, Vol. 2B, pp. 511–514.

[7]   Tarvas, S., and A. Isohatala, "An Internal Dual-Band Mobile Phone Antenna," *Proc. IEEE Antennas and Propagation Society Int. Symposium*, 2000, Vol. 1, pp 266–269.

[8]   Boyle, K. R., and P.J. Massey, "Nine-Band Antenna System for Mobile Phones," *Electronics Letters*, Vol. 42, No. 5, March 2006, pp. 265–266.

CHAPTER 3

Multiband Planar Wire Antennas

Manos M. Tentzeris, Bo Pan, and RongLin Li

3.1 Introduction

Wire antennas are the oldest and in many cases the most versatile antennas from a practical point of view [1]. In modern personal wireless applications, it is desirable to integrate the antenna on a circuit board for low cost, low profile, and conformality. A simple way to integrate a wire antenna with a circuit board is to print the wire on a planar printed circuit board (PCB), thus forming an integrated planar wire antenna. The manufacturing of planar wire antennas is much easier and more cost-effective than that of planar inverted-F antennas (PIFAs) which usually have a three-dimensional (3-D) configuration (see Chapter 1). The simplest planar wire configuration is a printed monopole. Figure 3.1 shows a single planar monopole above a ground plane. The monopole is fed by a microstrip line. Both of the monopole with a feeding line and the ground plane are integrated on a PCB. For a multiband operation, a multielement monopole may be needed. Figure 3.2 shows a two-element monopole for dual-band operation [2]. The dual-band monopole antenna consists of two monopoles with different lengths. The longer one (i.e., Monopole 1) is for a lower frequency band while the shorter one (i.e., Monopole 2) is dominant at the higher frequency. The two monopoles are combined at their lower ends and fed by a single microstrip line. A triple-band operation can be achieved by adding one more element to the dual-band monopole antenna.

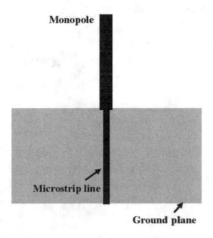

Figure 3.1 A planar monopole antenna.

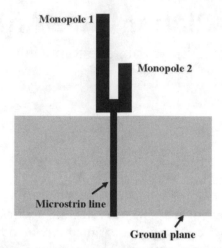

Figure 3.2   A dual-band planar monopole antenna.

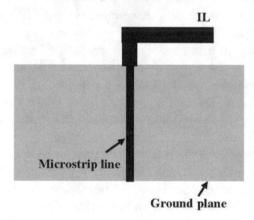

Figure 3.3   An inverted-L (IL) antenna.

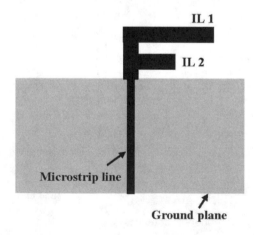

Figure 3.4   A double inverted-L antenna for dual-band operation.

Inverted L and inverted F antennas that have been introduced in Chapter 1 can also be implemented in a printed planar wire form (see Figures 3.3–3.6).

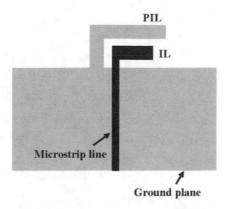

Figure 3.5 An inverted-L antenna with a parasitic inverted-L (PIL) element for bandwidth enhancement.

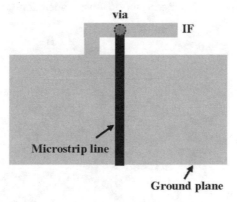

Figure 3.6 An inverted-F antenna.

Note that the inverted-L element is printed on the same side with the ground plane. Therefore, a metal via may be needed for connection between the inverted-L element and the microstrip line. The major advantage of this feeding configuration is that it increases the freedoms for impedance matching. By adjusting the position of the feeding point, it is easy to get good impedance matching for certain characteristic impedance.

By combining an inverted-F element with an inverted-L element, an inverted-FL antenna can be obtained for dual-band operation (see Figure 3.7). The inverted-L element decides the resonance of the lower frequency band whereas the inverted-F controls the impedance matching at the higher frequency band.

The impedance matching of an inverted-FL antenna can be improved by introducing a parasitic inverted-L (PIL) element near by the inverted-FL element (see Figure 3.8). By adjusting the shape of the PIL element, the impedance matching can be significantly improved [3].

When the two elements of a dual-band monopole antenna are folded in different sides (e.g., the longer one in the right side and the shorter one in the left side of the feeding point), we can get an asymmetric T-shaped antenna (see Figure 3.9). Therefore, the T-shaped antenna can be considered as a combination of two inverted-L elements. By adjusting the position of the feeding point (or the lengths of the T-shaped element), a dual-band operation can be achieved [4].

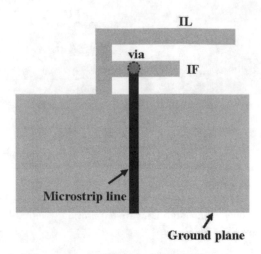

Figure 3.7   An inverted-FL antenna for dual-band operation.

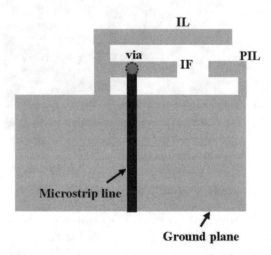

Figure 3.8   An inverted-FL antenna with a parasitic inverted-L element for bandwidth enhancement.

The bandwidth at the higher frequency band of a dual-band T-shaped antenna can increased by adding a parasitic inverted-L element below the shorter side of the T-shaped element (see Figure 3.10).

The length of an inverted-F antenna can be reduced by spiraling the horizontal segment of the inverted-F element (see Figure 3.11) [5].

The profile of a monopole antenna can also be lowered by folding the monopole into more complicated configurations, such as an S-shaped structure reported in [6, 7] (see Figure 3.12) or a meandered structure (see Figure 3.13). For a meandered structure, the dual-band operation can be achieved by combining with other configurations, such as an inverted-L element or forked branches.

A multiband operation can be also realized by a broadband antenna. An M-shaped monopole antenna with a T-shaped parasitic element on the ground plane (see Figure 3.14) has a compact configuration and can achieve a very wide band-

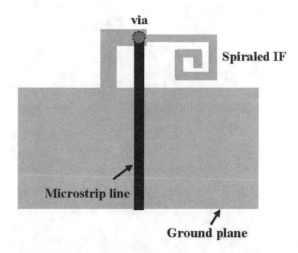

Figure 3.9   A spiraled inverted-F antenna.

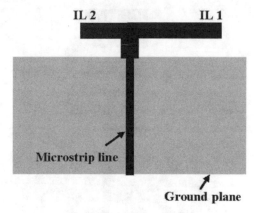

Figure 3.10   A T-shaped antenna for dual-band operation.

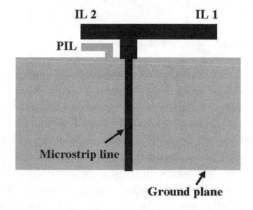

Figure 3.11   A T-shaped antenna with a parasitic inverted-L element for bandwidth improvement.

width, covering the frequency bands for GSM, GPS, DCS, PCS, UMTS, and WLAN [8].

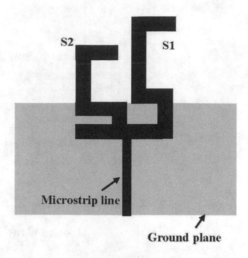

Figure 3.12   A double S-shaped antenna for dual-band operation.

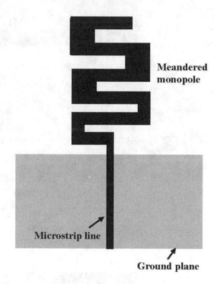

Figure 3.13   A meandered monopole antenna.

For a compact low-profile multiband antenna, it is usually difficult to achieve a wide bandwidth at the lower frequency band due to a smaller electrical size. A two-strip planar antenna (see Figure 3.15) can be used to overcome this difficulty [9]. This antenna consists of an S-strip and T-strip. The T-strip is directly fed by a microstrip line while the S-strip is electromagnetically coupled to the T-strip. The profile of the antenna is reduced through a folded configuration of the two stripes while the bandwidth is enhanced by the electromagnetic coupling between the T-strip and the S-strip. Multiband operation can be achieved by combining the two-strip antenna with other elements, such as planar monopoles [10].

In this chapter, we will introduce eight types of the multiband planar wire antennas. First, the dual-band monopole antennas will be presented. In Section 3.3, inverted-L antennas will be discussed. Inverted-F antenna and T-shaped will be investigated in Section 3.4 and 3.5, respectively. Two meandered antennas will

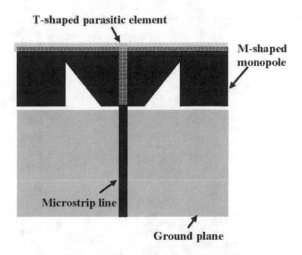

Figure 3.14   An M-shaped monopole antenna with a T-shaped parasitic element for broadband operation.

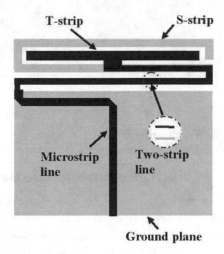

Figure 3.15   A two-strip broadband antenna.

be presented in Section 3.6. The M-shaped planar antenna will be given in Section 3.7. Finally, the two-strip antennas are developed for broadband and dual-frequency operations.

3.2   Multiband Planar Monopole Antennas

In this section, we will introduce three planar monopole antennas for dual-band operations: a microstrip-fed dual-band monopole antenna, a coplanar waveguide (CPW) fed dual-band monopole antenna, and a meandered CPW-fed multiband monopole antenna.

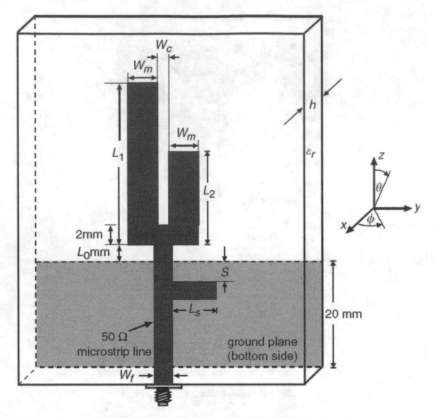

Figure 3.16 A realistic configuration of the microstrip-fed dual-band planar monopole antenna. (From: [2]. ©2001 IEEE. Reprinted with permission.)

3.2.1 Microstrip-Fed Dual-Band Monopole Antenna

The configuration of the microstrip-fed dual-band monopole antenna is shown in Figure 3.16 [2]. Two printed strip monopoles of different lengths, printed on the same side of an electrically thin substrate, are connected through a series microstrip line with a tuning stub. It is noted that the presence of the dielectric substrate lowers the fundamental resonance frequency of the antenna with air substrate and can provide mechanical support. The truncated ground plane on the backside of the substrate is used as reflector element. The tuning stub length was found to be very effective in controlling the coupling of the electromagnetic energy from the microstrip feed line to the strip monopole antenna, and good impedance matching for the dual-band antenna can be obtained.

The antenna was fabricated, tested, and compared with simulated results [2]. In Figure 3.17, the simulated values of input return loss of the final design are compared with measured data, showing good agreement. It can also be seen that the optimal antenna has more than 9% bandwidth at two design frequency bands.

3.2.2 CPW-Fed Dual-Band Monopole Antenna

The dual-band planar monopole can be also fed by a CPW line. CPW-fed antennas have many attractive features, such as no soldering points, easy fabrication and integration with monolithic microwave integrated circuits, and a simplified

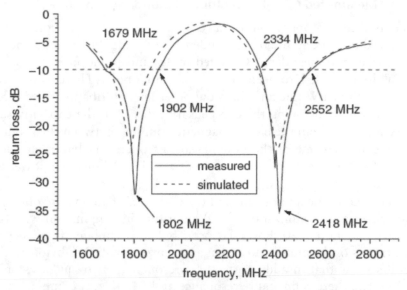

Figure 3.17 Measured and simulated results for the return loss of the optimized uniplanar monopole antenna. (From: [2]. ©2001 IEEE. Reprinted with permission.)

configuration with a single metallic layer. The operational principle of the CPW-
fed dual-band monopole antenna is similar to that of the microstrip-fed dual-
band monopole antenna; that is, to make use of a combination of two monopoles
connected in parallel at the feed point, each of the monopoles operating at a specified frequency band. Figure 3.18 shows the geometry of the CPW-fed dual-band
monopole antenna [11]. More details of this CPW-fed dual-band antenna can be
found in [11] and are not repeated here due to its similarity with the previously
discussed microstrip-fed monopole.

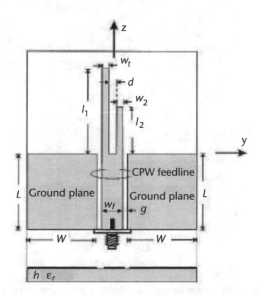

Figure 3.18 Geometry of a CPW-fed dual-band monopole antenna. (From: [11]. ©2004 IEEE.
Reprinted with permission.)

3.2.3 Meandered CPW-Fed Multiband Monopole Antenna

The multiband operation of a CPW-fed planar monopole antenna can be achieved by making use of a meandered CPW line. Figure 3.19 shows the configuration of the meandered CPW-fed multiband monopole antenna [12]. A 50Ω CPW line is used to feed the planar monopole. There is a meandered line between the CPW line and the monopole. The multiband impedance matching is obtained by controlling the compensation between the capacitive and inductive effects caused from the electromagnetic coupling between the ground planes and the meandered line at the desired various operating bands. In fact, the meandered CPW line acts as a band-pass filter, which allows the power to pass at the desired operating bands.

The geometrical parameters of the multiband antenna are obtained by using a particle swarm optimization (PSO) algorithm in conjunction with the method of moments. Figure 3.20 shows the measured and simulated frequency responses for return loss of the multiband antenna optimized by PSO. Obviously, the simulation results show that in addition to three resonances at frequencies of 1.9, 2.5, and 5.25 GHz, there is one more resonance at 4.04 GHz. A combination of two resonances at 1.9 and 2.5 GHz produces continuous bandwidth from 1.77 to 2.78

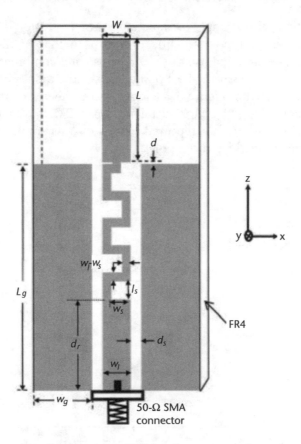

Figure 3.19 Geometry of a meandered CPW-fed planar monopole antenna for multiband operation. (From: [12]. ©2005 IEEE. Reprinted with permission.)

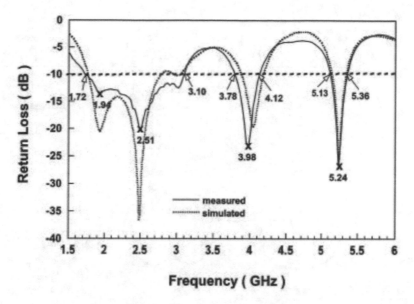

Figure 3.20   Measured and simulated results for return loss of the meandered CPW-fed multi-band monopole antenna (w=6.5 mm, L=26.88 mm, d=0.5 mm, d$_s$=2.53 mm, w$_g$=13.49 mm, L$_g$=49.4 mm, l$_s$= 3.96 mm, w$_s$=4.46 mm, and d$_f$=20 mm). (From: [12]. ©2005 IEEE. Reprinted with permission.)

GHz. The measured result shows four resonances at 1.94, 2.51, 3.98, and 5.24 GHz. The measured bandwidths are found to be about 1,380 MHz (1.72–3.1 GHz), 340 MHz (3.78–4.12 GHz), and 230 MHz (5.13–5.36 GHz), corresponding to relative bandwidths of 55%, 8.5%, and 4.4% at the center frequencies of 2.51, 3.98, and 5.24 GHz, respectively.

The measured radiation patterns at 1.94, 2.51, 3.98, and 5.24 GHz are plotted in Figure 3.21. Nearly omn-directional radiation pattern is observed in the x-y plane. The ranges of antenna gain are about 2.5–3.2 dBi, 3.2–4.1 dBi, and 4.0–4.3 dBi across the 2.51-, 3.98-, and 5.24-GHz bands, respectively.

Tri-band operation can be achieved by using additional arms and one design example can be found in [13].

3.3   Inverted-L Multiband Antennas

The profile of a monopole antenna can be reduced by folding it into an inverted-L shape. In this section, we will introduce a modified double inverted-L antenna and an inverted-L antenna with a parasitic element.

3.3.1   Modified Double Inverted-L Antenna for Multiband Operation

In [14], a low-profile planar monopole antenna was developed for multiband operation. The geometry and dimensions of the antenna are shown in Figure 3.22. This antenna can be considered as a modified double inverted-L antenna. By comparing the low-profile planar monopole with a double inverted-L antenna (Figure 3.3), we can see that the main difference between the two antenna configurations is the longer inverted-L (IL 1) element that is folded in the low-profile

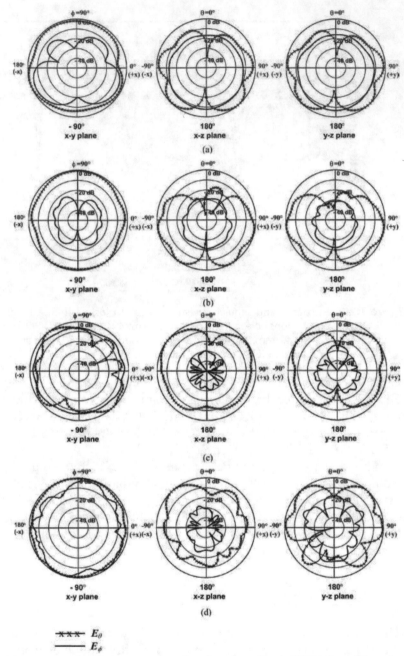

Figure 3.21 Measured radiation patterns for the meandered CPW-fed multiband monopole antenna (w=6.5 mm, L=26.88 mm, d=0.5 mm, d_s=2.53 mm, w_g=13.49 mm, L_g=49.4 mm, l_s= 3.96 mm, w_s=4.46 mm, and d_f=20 mm). (a) f =1.94 GHz, (b) f = 2.51 GHz, (c) f = 3.98 GHz, and (d) f = 5.24 GHz. (From: [12]. ©2005 IEEE. Reprinted with permission.)

planar monopole for a more compact configuration. The shorter inverted-L (IL 2) element is going from the feed point to the end of the inner strip encircled by the folded IL 1 element. The length of IL 1 is much greater than the length of IL 2, which makes the first resonant frequency much lower than the second reso-

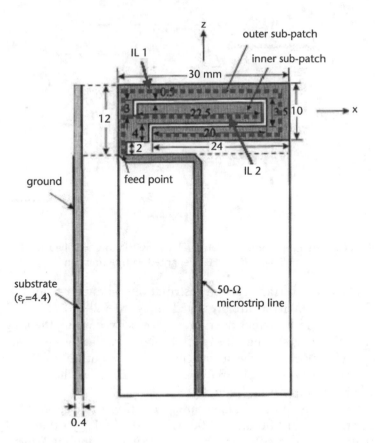

Figure 3.22 Geometry and dimensions of a modified double inverted-L antenna for multiband
operation. (From: [14]. ©2003 IEEE. Reprinted with permission.)

nance. To cover the 900-MHz GSM band, the length of IL 1 is about 70 mm,
which is slightly less than one-quarter wavelength of the operating frequency at
900 MHz. This difference is largely due to the effect of the supporting FR4 substrate, which reduces the resonant length of the radiating element. For the 1,800-
MHz DCS band, 1,900-MHz PCS band, and the 2,000-MHz UMTS band, the
length of IL 2 is chosen to about 30 mm, which makes it possible for the excitation of a quarter-wavelength resonant mode at about 2,000 MHz. This resonant
mode incorporating the second-higher (half-wavelength) resonant mode of the
longer inverted-L element, which is expected to be at about 1,800 MHz, forms a
wide impedance bandwidth covering the bandwidths of the 1,800-, 1,900-, and
2,000-MHz bands [14].

Figure 3.23 shows the measured return loss of the modified double inverted-
L antenna. It is seen that two operating bandwidths are obtained. The lower
bandwidth, determined by 1:2.5 VSWR, reaches 142 MHz and covers the GSM
band (890–960 MHz). The upper band has a bandwidth as large as 565 MHz
and covers the DCS (1,710–1,880 MHz), PCS (1,850–1,990 MHz), and UMTS
(1,920–2,170 MHz) bands. The measured return loss agrees with the simulated
(using IE3D) result.

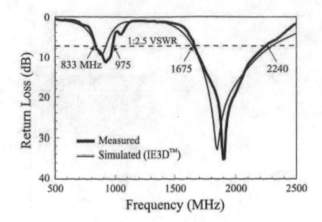

Figure 3.23 Measured and simulated return loss of the modified double inverted-L antenna. (From: [14]. ©2003 IEEE. Reprinted with permission.)

The excited surface current distributions obtained from the IE3D simulation at 900 and 1,900 MHz are presented in Figure 3.24. For the 900-MHz excitation, a stronger surface current distribution is observed along the longer inverted-L element. This suggests that IL 1 is the major radiating element at the 900-MHz band. At 1,900 MHz, the surface current distribution on IL 2 becomes stronger. This indicates that the shorter inverted-L element is the major radiating element for the higher operating frequency.

Figure 3.25 shows the measured radiation patterns at 900 and 1,800 MHz (the patterns at 1,900 and 2,000 MHz are similar to that at 1,800 MHz). The inverted-L antenna still has a radiation pattern roughly similar to that of a conventional monopole. The peak values of gain are 2.9, 3.0, 3.4, and 3.4 dBi in the 900-, 1,800-, 1,900-, and 2,050-MHz bands, respectively, and the gain variations are less than 1.5 dB.

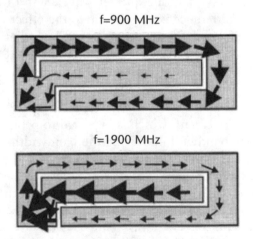

Figure 3.24 Simulated surface current distributions on the inverted-L elements at 900 and 1,900 MHz for the modified double inverted-L antenna. (From: [14]. ©2003 IEEE. Reprinted with permission.)

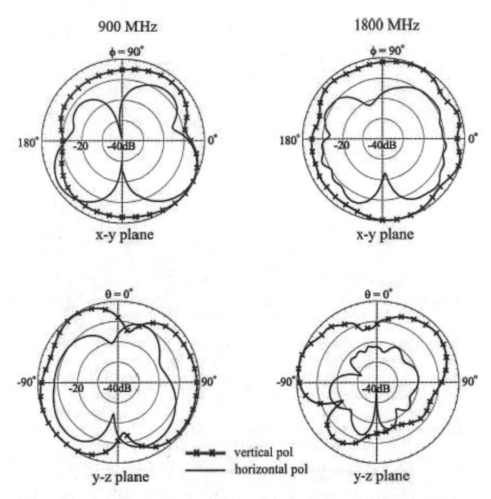

Figure 3.25 Measured radiation patterns at 900 and 1,800 MHz for the modified double inverted-L antenna. (From: [14]. ©2003 IEEE. Reprinted with permission.)

3.3.2 Inverted-L Antenna with a Parasitic Element for Dual-Band Operation

A single inverted-L element may generate one fundamental resonance. By adding a parasitic inverted-L element, a dual-band operation can be achieved. The geometry of an inverted-L planar antenna with a parasitic inverted-L element is shown in Figure 3.26 [15]. This antenna was designed for wireless applications in the 2.4-, 5.2-, and 5.8-GHz bands. The inverted-L element is used as a driven element that is connected to a 50Ω microstrip line. The driven element and the parasitic element are printed on the same side of an FR4 substrate. The parasitic element is connected to the ground plane by a shorting pin. The parasitic inverted-L element has a horizontal length l_1=12.5 mm and a vertical height h_1=8 mm, while the driven inverted- L element has a horizontal length l_2=7 mm and a vertical height h_2=6 mm. The parasitic inverted-L element is closely placed on the top of the driven inverted-L element. The distances between them are 1 mm in length at the top and S=2 mm at the right, respectively. The driven inverted-L element can resonate as a monopole antenna with the total length (l_2+h_2) determined by nearly

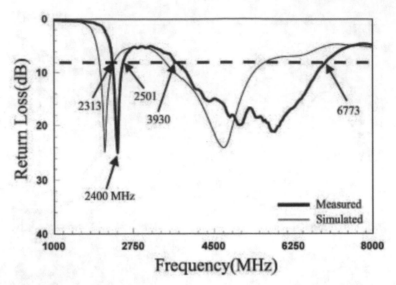

Figure 3.26 Geometry of an inverted-L antenna with a parasitic inverted-L element for dual-band operation. (From: [15]. ©2004 IEEE. Reprinted with permission.)

one-quarter wavelength of the upper operating frequency. Probably owing to the coupling of the driven element and the parasitic element, the resonant length of the parasitic inverted-L element is reduced. The total length (l_1+h_1) is found to be about two thirds of one quarter-wavelength of the lower operating frequency.

Figure 3.27 shows the measured and simulated (using the simulation software High Frequency Structure Simulator—HFSS) results for return loss of the dual-band inverted-L antenna. It is seen that two operating bandwidths are obtained. The lower bandwidth reaches 188 MHz and covers the 2.4-GHz WLAN band (2,400–2,484 MHz). The upper band has a bandwidth as large as 2,843 MHz and covers other WLAN bands at 5.2 GHz (5,150–5,350 MHz) and 5.8 GHz (5,725–5,852 MHz).

3.4 Inverted-F Antennas for Dual-Band Operation

Inverted-F antennas have a low profile and are easy to match with difference impedance by adjusting the position of feeding point. A dual-band operation for an inverted-F antenna can be realized by folding the inverted-F element or by combining with other elements, such as an inverted-L element. In this section, we will introduce a spiraled inverted-F antenna, a coupling inverted-F antenna, and an inverted-F antenna combined with an inverted-L element (called an inverted-FL antenna).

3.4.1 Spiraled Inverted-F Antenna for Dual-Band Operation

An inverted-F antenna can realize dual-band operation by loading the conventional inverted-F antenna with a spiral inductor (called a spiraled inverted-F antenna) [5]. The spiral loading also miniaturizes the size of a convention

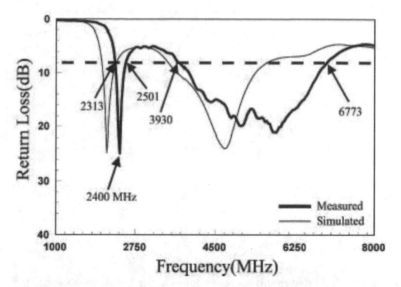

Figure 3.27 Measured and simulated results for return loss of the inverted-L antenna with a parasitic inverted-L element (l_1=12.5 mm, S=2 mm; h_1=8 mm, l_2=7 mm, h_2=6 mm). (From: [15]. ©2004 IEEE. Reprinted with permission.)

inverted-F antenna. Figure 3.28 shows how the spiraled inverted-F antenna is evolved from a conventional inverted-F antenna.

The four inverted-F antennas presented in Figure 3.28 were simulated using IE3D. The total length of the inverted-F antennas from the short point to the open end is 25.5 mm. Figure 3.29 shows the simulated results for return loss of the four inverted-F antennas. It is seen that the fundamental resonant frequency (i.e., the

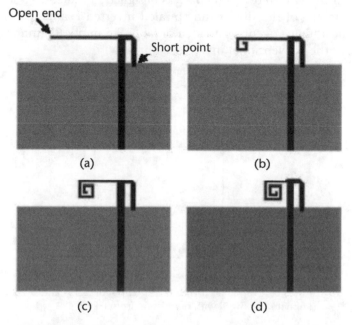

Figure 3.28 (a–d) Evaluation of a spiraled inverted-F antenna. (From: [15]. ©2007 IEEE. Reprinted with permission.)

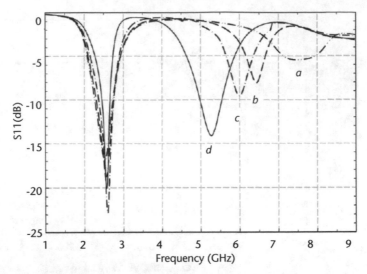

Figure 3.29 Simulated results for return losses of the inverted-F antennas shown in Figure 3.28 (substrate thickness=0.8 mm, ε_r=4.4, ground plane= 46 mm × 55 mm). (From: [5]. ©2007 IEEE. Reprinted with permission.)

first resonance) of these antennas remains about the same. As expected, the bandwidth is decreased as the inverted-F element is spiraled due to the increase in Q factor of the spiraled antenna. However, the frequency at the second resonance becomes lower as the inverted-F antenna is spiraled. The second resonant frequency reduces by about 71% (from 7.35 to 5.25 GHz) when the antenna structure is changed from the conventional inverted-F antenna [Figure 3.28(a)] to the spiraled inverted-F antenna [Figure 3.28(d)].

A spiraled inverted-F antenna was designed in the 2.45- and 5.25-GHz bands. The geometry of the dual-band spiraled inverted-F antenna is shown in Figure 3.30. The antenna occupies an area of $l \times h$=9.5 mm × 6.5 mm, which is only 50% of that of the conventional inverted-F antenna.

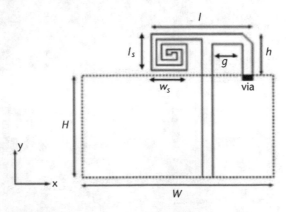

Figure 3.30 A dual-band spiraled inverted-F antenna for 2.45- and 5.25-GHz frequency bands. (Reprinted from [5] with permission from IEEE.)

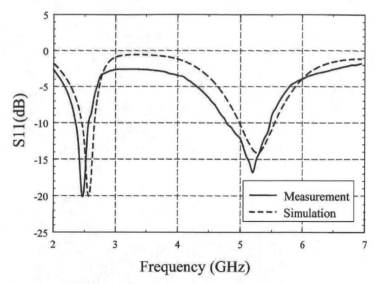

Figure 3.31 Measured and simulated results for return loss of the dual-band spiraled inverted-F antenna (substrate thickness=0.8 mm, ε_r=4.4, l=9.5 mm, h=6.5 mm, W=46 mm, H=55 mm, l_s=4.5 mm, w_s=4 mm, g=1.54 mm, the width and gap of the spiral=0.5 mm). (From: [5]. ©2007 IEEE. Reprinted with permission.)

Figure 3.31 shows the measured return loss compared with the simulated result. The measured 10-dB bandwidth is 140 MHz at 2.45-GHz band and 756 MHz at 5.25-GHz band.

The radiation patterns measured at 2.45 and 5.25 GHz are presented in Figure 3.32. The radiation pattern at the lower frequency is similar to that of a conventional inverted-F antenna. The peak values of the antenna gain arc 1.6 dBi and 3.0 dBi at 2.45 GHz and 5.25 GHz, respectively.

3.4.2 Coupling Dual-Band Inverted-F Antenna

The size of a conventional inverted-F antenna can be also reduced by introducing a coupling configuration as illustrated in Figure 3.33 [5]. The size of the conventional inverted-F antenna is limited by the large inverted-L structure. This length can be reduced by cutting part of the inverted-L element and then connecting a shunt capacitor with suitable capacitance. Instead of a lumped capacitor, a printed capacitor is used to accomplish the goal in Figure 3.33. First let the end of the large inverted-L connect to the top side of the substrate by a via and bend downwards. Then a small strip is introduced from the ground plane on the bottom to make a gap or overlap region with the downwards bending end. This produces an equivalent capacitor with one terminal connected to the end of the inverted-L and the other to the ground. The capacitance can be adjusted by changing the size of the gap or overlap region. Instead of a direct contact, the feeding microstrip line is connected to the horizontal metal line through a coupling section of length l_2. The structure can also be viewed as two coupled monopole antennas with different orientation. The small inverted-L monopole antenna on the top side of the substrate is designed for operation at the higher frequency, whereas the large monopole antenna on the bottom is for the lower frequency.

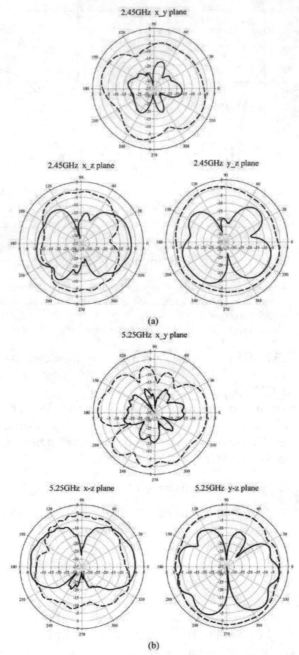

Figure 3.32 Measured radiation patterns for the dual-band spiraled inverted-F antenna at (a) 2.45 GHz and (b) 5.25 GHz (solid line denotes E-theta; dashed line denotes E-phi). (From: [5]. ©2007 IEEE. Reprinted with permission.)

The complete structure of the antenna occupies a region of $l \times h$=12 mm × 4.9 mm, which is about 50% of a conventional inverted-F antenna. The small inverted-L antenna has a total length ($h+l_2$) about a quarter wavelength at 5.25 GHz while the large one has a length ($l+l_b+h$) near a quarter wavelength at 2.45 GHz. The length of the coupling section can be adjusted to get good impedance

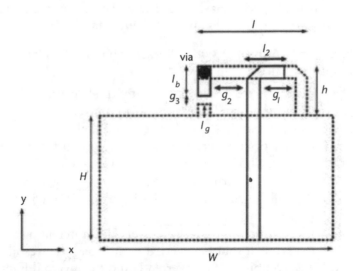

Figure 3.33   A dual-band coupling inverted-F antenna. (From: [5]. ©2007 IEEE. Reprinted with permission.)

matching at both bands. The other lengths (l_b and h) are changed according to the required resonant lengths.

Figure 3.34 presents the measured return loss compared with the simulation result. The measured 10-dB bandwidth around 2.45 GHz is 240 MHz while the

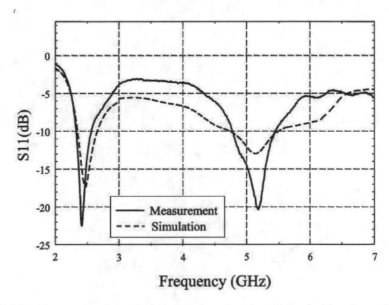

Figure 3.34   Measured and simulated results for return loss of the dual-band coupling inverted-F antenna (substrate thickness=0.8 mm, ε_r=4.4, l=12 mm, h=4.9 mm, W=46 mm, H=55 mm, l_2=4.8 mm, l_b=2.2 mm, l_g=1.2 mm, g_1 =4.2 mm, g_2=4.4 mm, g_3 =1.2 mm, the strip width of the large monopole=1.0 mm). (From: [5]. ©2007 IEEE. Reprinted with permission.)

bandwidth around 5.25 GHz is 672 MHz. The gain and radiation pattern are similar to those of the spiraled inverted-F antenna.

3.4.3 Inverted-FL Antennas with a Parasitic Inverted-L Element for Bandwidth Enhancement

By combining an inverted-F with an inverted-L element, an inverted-FL antenna can be formed for dual-band operation. To improve the impedance matching at the higher frequency band, a parasitic inverted-L element is needed. Figure 3.35 shows the geometry of an inverted FL antenna with different parasitic inverted-L elements [3]. Figure 3.35(b) is a modified version of the structure in Figure 3.35(a), where a small protrusion of area is added to the parasitic element in Figure 3.35(a). As will be shown later, the inverted-FL antenna in Figure 3.35(b) achieves better performance.

As shown in Figure 3.36, a longer inverted-L antenna has a resonance at the lower frequency while a shorter inverted-F antenna has a resonance at the higher frequency. When combining the inverted L and the inverted F into an inverted-FL

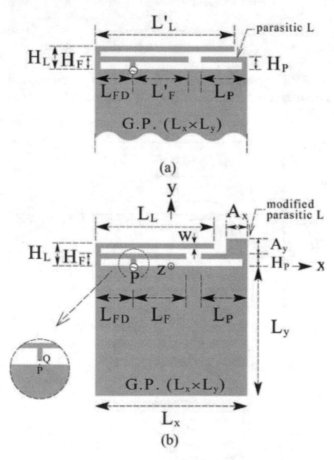

Figure 3.35 Geometry of dual-band inverted-FL antennas. (a) Inverted-FL antenna with a parasitic inverted-L element, and (b) inverted-FL antenna with a modified parasitic inverted-L element. (From: [3]. ©2005 IEEE. Reprinted with permission.)

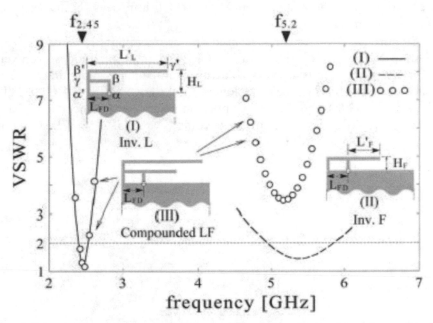

Figure 3.36   Simulated results for VSWR of an inverted-L antenna (H_L=4.5 mm, L'_L=27.5 mm, L_{FD}=7.5 mm), an inverted-F antenna (H_F=2.5 mm, L'_F=11 mm, L_{FD}=7.5 mm), and an inverted-FL that is a combination of the inverted L and the inverted F. (From: [3]. ©2005 IEEE. Reprinted with permission.)

antenna, two resonances appear. However, the impedance matching at the higher frequency becomes worse. To solve this problem, a parasitic inverted-L element can be added to the inverted-FL antenna.

3.5   T-Shaped Antennas for Multiband Operation

A T-shaped planar monopole can be considered as a combination of two inverted-L monopoles. By adjusting the lengths of two inverted-L monopoles properly, it is possible to achieve a dual-band operation. In this section, we will introduce three T-shaped antennas: a single T-shaped antenna, a double T-shaped antenna, and a single T-shaped antenna with a parasitic inverted-L element for bandwidth improvement.

3.5.1   Single T-Shaped Antenna for Multiband Operation

Figure 3.37 shows the geometry and dimensions of a single T-shaped antenna for multiband operation [4]. The antenna is designed for 2.4/5-GHz WLAN bands. The radiating element has compact dimensions of 10 mm × 26 mm. Two asymmetric horizontal strips on the top of the T-shaped antenna can produce two different surface current paths and result in a dual-resonance mode. The right horizontal strip (the longer one) provides the lower mode covering the 2.4-GHz WLAN band, while the left one (the shorter one) controls the upper mode including 5-GHz WLAN bands. A conducting triangular section is also added into the vertical strip of the T-shaped antenna, which improves the impedance matching for both bands.

Figure 3.38 shows the measured and two simulated results for VSWR of the T-shaped antenna. Two separate broadband resonant modes are excited at 2.4 and 5.5 GHz simultaneously with good impedance matching. The impedance bandwidth (VSWR<2) of the lower mode covers the 2.4-GHz WLAN band (2,400–2,484 MHz). For the upper mode, a wide impedance bandwidth is obtained, which is sufficient to encompass the 5-GHz WLAN band (5,150–5,350/5,470–5,825 MHz).

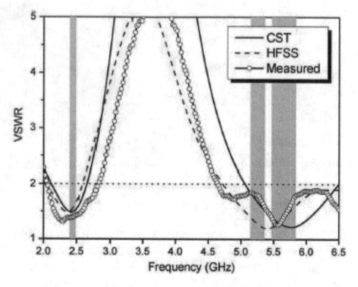

Figure 3.37 Measured and simulated results for VSWR of the T-shaped monopole antenna. (From: [4]. ©2006 IEEE. Reprinted with permission.)

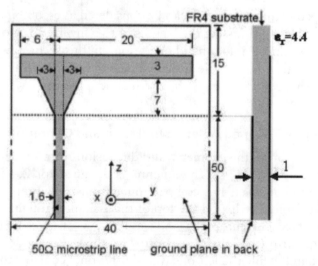

Figure 3.38 Geometry and dimensions of a T-shaped monopole antenna (unit: mm). (From: [4]. ©2006 IEEE. Reprinted with permission.)

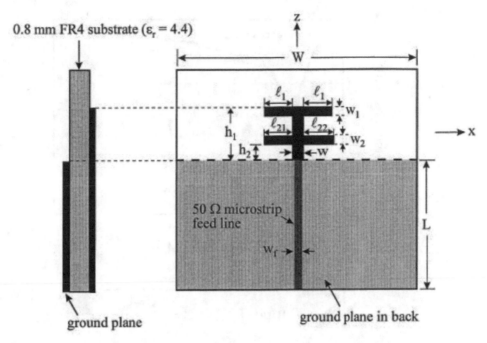

Figure 3.39 Geometry of a double-T monopole antenna. (From: [16]. ©2003 IEEE. Reprinted with permission.)

3.5.2 Double T-Shaped Antenna for Dual-Band Operation

A dual-band operation can be also achieved by a double T-shaped planar antenna. Figure 3.39 shows the geometry of a double-T shaped antenna that consists of two stacked T-shaped monopoles [16]. Both of the two T-shaped monopoles are printed on the same side of an FR4 dielectric substrate. The larger T-shaped monopole comprises a vertical strip (width w and height h_1) in the center and a horizontal strip (width w_1 and length $2l_1+w$) on the top. The larger T-shaped monopole controls the first (i.e., the lower) operating band. On the other hand, a lower horizontal strip of width w_2 and length ($l_{21}+w+l_{22}$) and a portion (length=h_2) of the vertical strip form the smaller T-shaped monopole, which controls the second (i.e., the upper) operating band. To achieve the desired dual-band operations, for example, the WLAN operations in the 2.4/5.2-GHz bands, lengths and widths of the larger and smaller T-shaped monopoles are adjusted for controlling, respectively, the lower and upper operating bands. The optimal parameters were obtained using HFSS.

To demonstrate the dual-band operation of the double T-shaped antenna, a reference antenna (i.e., the antenna shown in Figure 3.39 without the lower horizontal strip) was first tested. The measured and simulated results for return loss of the reference antenna are shown in Figure 3.40(a). It is seen that only one resonance is excited at the 2.4-GHz band. By adding a lower horizontal strip to the reference antenna, an additional resonance at the 5.2-GHz band can be obtained. The measured and simulated results for return loss of the double T-shaped antenna are shown in Figure 3.43(b). The bandwidths of the double T-shaped antenna are 390 MHz (2,175–2,565 MHz) at the 2.4-GHz band and 105 MHz

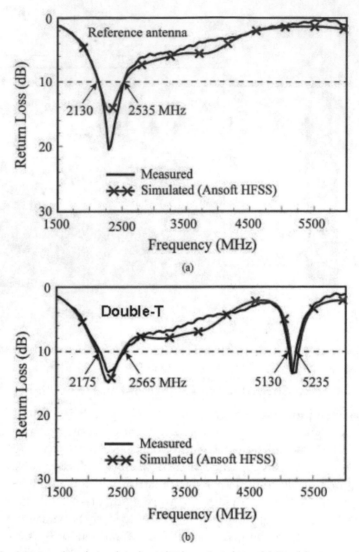

Figure 3.40 Measured and simulated results for return loss of (a) a reference antenna and (b) a double-T antenna (l_1=5.3 mm, h_1=14.5 mm, h_2= 5 mm, w=1.5 mm, w_1=1.5 mm, w_2 =1.5 mm, w_f=1.5 mm, L=50 mm, W=75 mm, l_{21}=l_{22}=7.3 mm). (From: [16]. ©2003 IEEE. Reprinted with permission.)

(5,130–5,235 MHz) at the 5.2-GHz band. The bandwidth at the 5.2-GHz band can be improved by increasing the strip widths (w, w_1, w_2) [16].

3.5.3 T-Shaped Antenna with a Parasitic Inverted-L Element for Bandwidth Improvement

The bandwidth of a T-shaped antenna can be increased by introducing a parasitic element. Figure 3.41 shows the geometry of a T-shaped dual-band antenna with a parasitic inverted-L element [17]. The lower section of the T-shaped antenna is folded to reduce the antenna profile. The antenna is mounted above the top edge of a large ground plane, which is considered here as the supporting metal frame for the display of a laptop. The antenna is fed by using a 50Ω coaxial line, with its

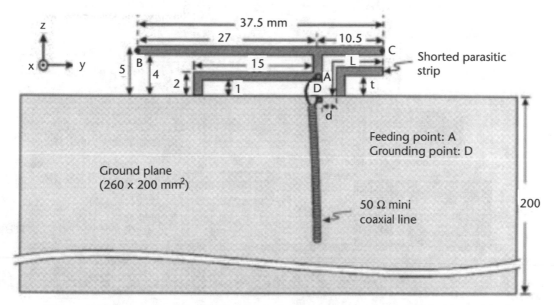

Figure 3.41 Geometry of a T-shaped monopole antenna with a parasitic inverted-L element. (From: [17]. ©2005 IEEE. Reprinted with permission.)

central conductor connected to point A (the feeding point) and its outer ground sheath connected to point D (the grounding point). The antenna was fabricated from line-cutting a 0.2-mm brass sheet and then connected to the ground plane. For practical applications, the proposed antenna and the ground plane (metal frame) together can also be fabricated by stamping a large metal plate; that is, the proposed antenna becomes an integrated antenna for laptop applications.

To demonstrate the effect of the parasitic element on the impedance bandwidth, Figure 3.42 shows the measured results for return loss of the T-shaped antenna with and without the parasitic element. The presence of the parasitic strip indeed greatly increases the antenna's upper bandwidth from 630 to 2,140 MHz. This bandwidth enhancement is mainly due to a second upper resonance generated at frequencies larger than 6 GHz. The second upper mode and the first upper mode excited at frequency lower than 5 GHz form the wide upper operating band.

Figure 3.43 shows the measured and simulated results for return loss of the T-shaped monopole antenna with a parasitic element. From the measured result, the lower band has an impedance bandwidth of 190 MHz (about 8%) at the 2.4-GHz band. The impedance bandwidth is 2,140 MHz (about 39%) at the 5.5-GHz band.

3.6 Meandered Multiband Antennas

Meandering is a well-known approach to antenna miniaturization. In this section, we will introduce two meandered planar antennas for multiband operation, including a nonuniform meandered antenna and a compact multiband meandered planar antenna.

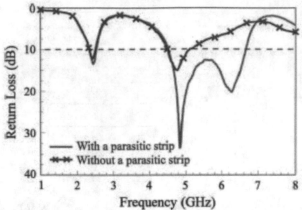

Figure 3.42 Measured results for return loss of a T-shaped monopole antenna with and without a parasitic element (d= 0.5 mm, t=2 mm, L=12 mm). (From: [17]. ©2005 IEEE. Reprinted with permission.)

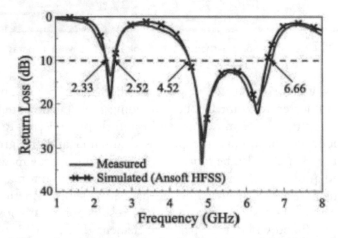

Figure 3.43 Measured and simulated results for return loss of the T-shaped monopole with a parasitic element (d= 0.5 mm, t=2 mm, L=12 mm). (From: [17]. ©2005 IEEE. Reprinted with permission.)

3.6.1 A Nonuniform Meandered Antenna with a Forked Ground Plane For Multiband Operation

The geometry of the multiband meandered monopole antenna is illustrated in Figure 3.44 [18]. This antenna was designed for IEEE802.11b/g/a triple-band WLAN applications using a 2.5D method of moments (MoM) full-wave simulator. A top patch and a bottom patch are printed on both sides of a substrate. The top patch is a nonuniform meander-line that is responsible for generating two resonant modes. The total length of the meander-line is about a quarter wavelength of the lower frequency resonant mode. The meander-line can be divided into two segments with different widths. The distance between the open end and the width-discontinuity point of the meander-line is about a quarter wavelength of the higher-frequency resonant mode. The bottom patch is a forked ground that is responsible for generating the third resonant mode and improving the impedance

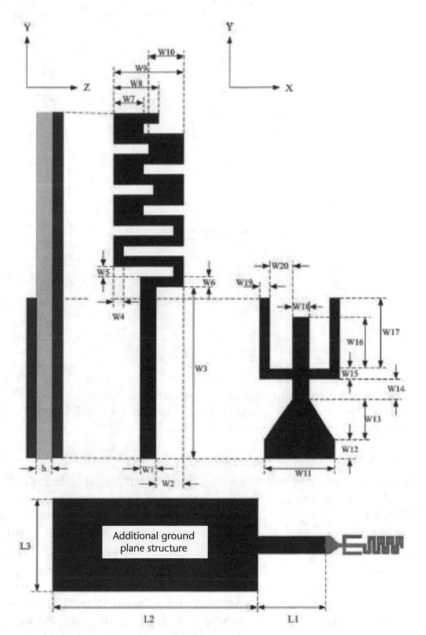

Figure 3.44 Geometry of a nonuniform meandered monopole antenna with a forked ground
plane for IEEE802.11b/g/a WLAN applications (W1 = W18 = 1:6 mm, W2 = 2:7 mm,
W3 = 17 mm, W4 = W5 = W6 = W15 = W19 = 1 mm, W7 = 3 mm, W8 = 4:5 mm,
W9 = W11 = 7 mm, W10 = 3:5 mm, W12 = 1:9 mm, W13 = 2:7 mm, W14 = 2 mm,
W16 = 5:1 mm, W17 = 7 mm, W20 = 2:2 mm, L1 = 30 mm, L2 = 90 mm, L3 = 40
mm, and h = 0:8 mm). (From: [18]. ©2006 IEEE. Reprinted with permission.)

bandwidth performance. The length W17 is about a quarter wavelength of the
third resonant mode. The lower end of the forked ground is gradually widened
and eventually connected to the ground plane of a 50Ω microstrip line.

The simulated and measured results for return loss of the meandered antenna
are shown in Figure 3.45. The measured impedance bandwidths (VSWR<2.5)

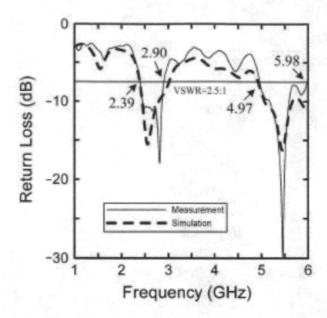

Figure 3.45 Measured and simulated results for return loss of the nonuniform meandered monopole antenna. (From: [18]. ©2006 IEEE. Reprinted with permission.)

meet the requirement for IEEE 802.11b/g/a applications. The radiation pattern of
the antenna is similar to that of a conventional monopole. The peak gains are
approximately 1.48, 2.30, and 3.05 dBi at the 2.4-, 5.2-, and 5.8-GHz bands,
respectively.

3.6.2 A Compact Multiband Meandered Planar Antenna

The geometry of a compact multiband meandered planar antenna is shown in Figure 3.46(a) [19]. The antenna was design for GSM-900, DCS-1800, PCS-1900,
UMTS-2000, and WLAN-2400 applications. The whole antenna occupies an area
of 38.5 mm × 15 mm. There are three meandered monopoles starting from the
point a, considered as three radiating elements or branches. The first branch (a-b-
c-d) and the second branch (a-b-c-e), 78- and 95-mm long, respectively, generate
the resonant mode for the GSM band. The length of the third branch (a-b-f) is
about 38 mm, approximately one-quarter wavelength of 2-GHz frequency. The
additional branch (g'-g) is for tuning and affects the higher modes of the first and
the second branches, resulting in a wide bandwidth for WLAN-2400 band. The
three radiating elements are not independent but coupling with each other, so that
they have to be optimized jointly to meet the multiband operation requirements.
The detailed dimensions of the antenna are given in Figure 3.46(b).

The measured and simulated (using HFSS) results for return loss of the
antenna are shown in Figure 3.47, from which good agreement can be observed.
It is seen that the first band is from 891 to 961 MHz for VSWR<2.5, which is
formed by two resonating modes close to each other and satisfies the GSM operation. The second band is from 1,705 to 2,180 MHz, covering the DCS, PCS, and
UMTS bands. The third band is from 2,341 to 2,980 MHz, sufficient for WLAN

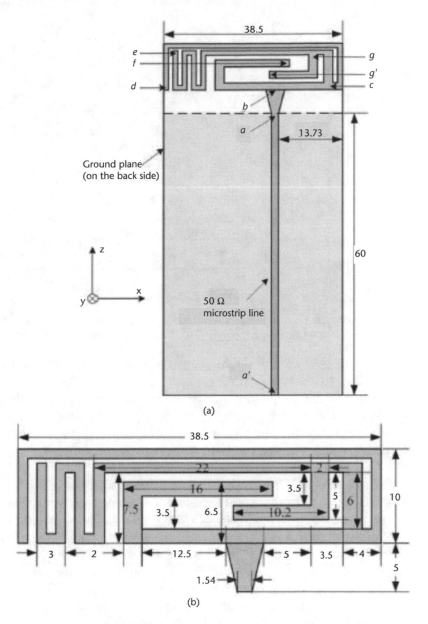

Figure 3.46 Geometry of a compact multiband meandered planar monopole. (a) Top view, and (b) detailed dimensions of radiating element (unit: mm). (From: [19]. ©2006 IEEE. Reprinted with permission.)

operation. The measured antenna gains at the three frequencies are 1.09, 1.64, and 3.19 dBi, respectively.

3.7 M-Shaped Planar Monopole for Multiband Operation

Multiband operation can be also achieved by using a wideband antenna. In this section, we will introduce a compact M-shaped planar antenna for multiband

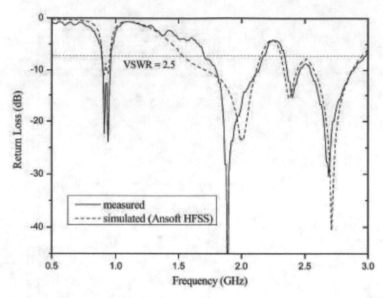

Figure 3.47 Measured and simulated results for return losses of the compact multiband mean-dered planar antenna. (From: [19]. ©2006 IEEE. Reprinted with permission.)

operation. The geometry of the M-shaped planar antenna is illustrated in Figure 3.48 [8]. The "M"-like radiating element consists of three parts: a rectangular patch (length 50 mm and width a), two trapezoidal patches with same parameters (height c, lower width d and upper width 9 mm), and an isosceles triangular patch (length g and height b). The "M"-like radiating element is printed on a substrate with thickness 0.4 mm and the relative dielectric constant ε_r=4.4. A 50Ω micros-trip line is used to feed the proposed antenna, and is printed on the same side of the substrate. On the other side of the substrate, there is a solid main ground plane below the microstrip feed line and a T-shaped ground plane protruding from the main ground plane below the "M"-like radiating patch. The T-shaped ground plane comprises a central vertical strip (length n and width 0.74 mm) and a top horizontal strip (length 50 mm and width m).

The geometrical parameters (a, b, c, d, g, m, and n) were optimized for the best bandwidth performance by numerical simulation using IE3D. It is found that the rectangular patch, whose width and length determine the main operating frequency band, is the dominating radiating patch. The isosceles triangular patch, whose apex angle and height determine the bandwidth, is used to expand the bandwidth. The T-shaped ground plane will also affect the bandwidth, but more importantly, it dominates the lowest operating frequency. The two trapezoidal patches can tune the bandwidth and the lowest operating frequency slightly. Another important point is that the "M"-like radiating element of the proposed antenna should be designed jointly with the ground plane to obtain an optimum antenna, because the ground plane with the specific T-shape is, in fact, a very important part of the antenna. Considering the relative bandwidth, the lowest operating frequency, and the dimensions of the proposed antenna, the optimum parameters are obtained by simulation, which are: a=20 mm, b=8 mm, c=6 mm, d=5 mm, g=32 mm, m=4 mm, and n=26 mm. Using the above-mentioned param-

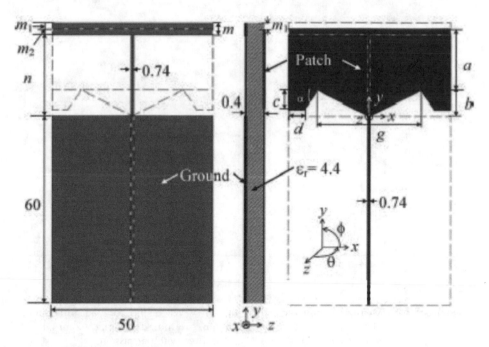

Figure 3.48 Geometry of a compact multiband M-shaped planar antenna (unit: mm) (From: [8]. ©2006 IEEE. Reprinted with permission.)

eters, a prototype of the proposed antenna was fabricated by the artwork of digital circuits and using a normal printed digital circuit board. The simulated and measured results for VSWR are shown in Figure 3.49. It is found that the bandwidth (VSWR<3) satisfies the requirements of the current and future mobile communication and navigation services such as GSM, GPS, DCS, PCS, UMTS, and WLAN. Furthermore, the height of the proposed antenna is less than 9% of the free-space wavelength of the lowest desired frequency at 890 MHz, so that it is fairly small for using in mobile terminals such as handsets.

3.8 Two-Strip Planar Antennas for Broadband and Multiband Operation

In the previous sections, it has been demonstrated that an S-shaped monopole or a T-shaped monopole can be used to lower the profile of a conventional monopole antenna. However, the bandwidth of the low-profile monopole antenna becomes narrow. In this section, we combine the S-shaped monopole with a T-shaped monopole to increase the bandwidth and to realize multiband operation.

3.8.1 Broadband Two-Strip Antenna

For GPS (1,570–1,580 MHz), DCS-1800 (1,710–1,880 MHz), IMT-2000 (1,885–2,200 MHz), and WLAN-IEEE 802.11b (2,400–2,483 MHz) applications, it is required to have a bandwidth of ~50%. In this section, we present a fully planar antenna that can achieve a bandwidth of more than 50% and has a more compact size (18 mm × 7.2 mm × 0.254 mm) than previous antennas. The configuration of

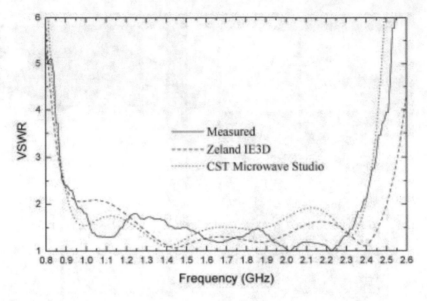

Figure 3.49   Measured and simulated results for VSWR of the compact multiband meandered planar antenna (a=20 mm, b=8 mm, c=6 mm, d=5 mm, g= 2 mm, m =4 mm, n=26 mm). (From: [8]. ©2006 IEEE. Reprinted with permission.)

the compact planar antenna is illustrated in Figure 3.50 [9]. The planar antenna consists of an S-strip and a T-strip that are printed on the two sides (i.e., the front side for T-strip and the backside for the S-strip) of the substrate, respectively. There is no direct electrical connection (e.g., by a shorting via) between the front side and the backside. The T-strip (its strip width=w_s) is fed by a 50Ω microstrip line while the S-strip (its strip width=w_s) is terminated at a ground plane (its length=L_g). The upper section (its width=W_T) of the T-strip is fitted (leaving a space of w_s) into an area surrounded by the upper section of the S-strip while the lower section of the T-strip overlaps with the lower section of the S-strip, forming a two-strip line. The height (H) of the planar antenna can be adjusted for an optimal performance. The compactness of the planar antenna is attributed to the folded configuration of the S-strip and T-strip (specifically, the folded two-strip line) while the broadband performance is a result of the mutual coupling between the S-strip and the T-strip.

The feeding point is set up at the center of the upper half section of the ground plane. This setup is completely for the purpose of accurate measurement. As will be shown soon, the length (L_g) of the ground plane can affect the performance of the antenna. If the feeding point is set up in the lower section of the ground plane, a coaxial cable attached to the ground plane tends to increase the length of the ground plane, hence changing the antenna performance. In realistic topologies of mobile handsets, the ground plane can be a part of a printed circuit board (PCB) and the planar antenna can be directly connected to the output of an RF front end (or an RF chip). Therefore this setup reflects a practical scenario where there is no coaxial cable attachment.

Figure 3.51 shows the simulation results for return loss at three different heights, namely, H=6.0, 7.2, and 8.4 mm. Note that the height H is adjusted by changing the number of the equidistant crossbars of the T-strip (e.g., 2, 3, and 4

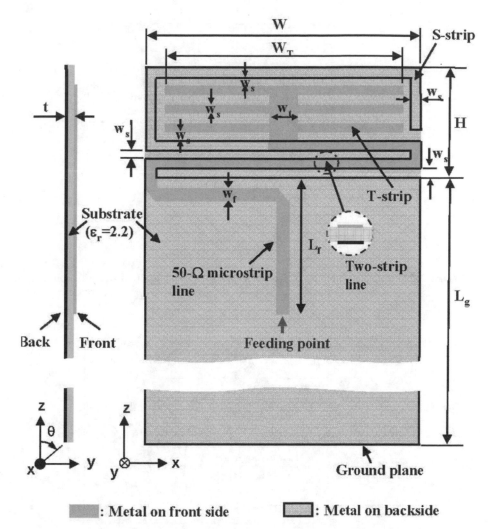

Figure 3.50   Configuration of a compact broadband planar antenna (H=7.2 mm, W=18 mm, W$_T$=15.6 mm, w$_s$=0.6 mm, w$_t$=1.8 mm, w$_f$=0.75 mm, t=0.254 mm L$_f$=15 mm). (From: [9]. ©2007 IEEE. Reprinted with permission.)

crossbars for H=6.0, 7.2, and 8.4 mm, respectively). It is observed from Figure 3.50 that the planar antenna has a maximum bandwidth at H=7.2 mm. The maximum bandwidth is close to 50%. The total size of the planar antenna is 18 mm × 7.2 mm × 0.254 mm, which is more compact than the previously published antennas.

The effect of the length (L$_g$) of the ground plane on the return loss of the planar antenna is exhibited in Figure 3.52. We can see that there is an optimum value for L$_g$ where the planar antenna has the best performance for return loss. The reason for the length dependence is that the ground plane also serves as a radiating element. The optimum value for L$_g$ is found to be around 60 mm. (Probably around L$_g$=60 mm, the ground plane, the S-strip, and the T-strip form a resonant radiator.) Therefore, the planar antenna is suitable for compact mobile handsets.

To verify the performance of the compact broadband planar antenna, a prototype was fabricated and measured. Four photographs of the antenna prototype

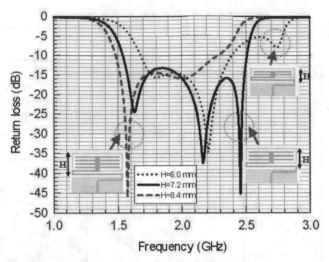

Figure 3.51   Simulated results for return loss of the planar antenna with different heights (H=6.0, 7.2, and 8.4 mm) (L_g=60 mm). (From: [9]. ©2007 IEEE. Reprinted with permission.)

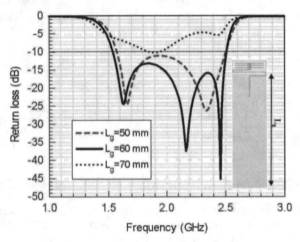

Figure 3.52   Effect of the length (L_g) of ground plane on the return loss of the compact broadband planar antenna (H=7.2 mm). (From: [9]. ©2007 IEEE. Reprinted with permission.)

are inset in Figure 3.53 to show the front view and the back view of the planar antenna. For the purpose of measurement, the antenna is connected to a 0.085″ semirigid coaxial cable in the upper section of the ground plane. After using this feeding structure, we found that the effect of the coaxial cable on the measurement results could be significantly alleviated. The measured return loss (RL) is presented in Figure 3.53, which shows a bandwidth of more than 50% for RL<-10 dB, covering the frequency bands for GPS (1,570–1,580 MHz), DCS-1800 (1,710–1,880 MHz), IMT-2000 (1,885–2,200 MHz), and WLAN-IEEE 802.11b (2,400–2,483 MHz).

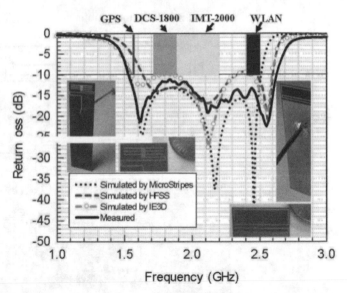

Figure 3.53 Measured return loss of the compact broadband planar antenna compared with simulation results. (Inset are four photographs of the antenna prototype, which show the front view and the back view of the planar antenna, respectively.) (From: [9]. ©2007 IEEE. Reprinted with permission.)

3.8.2 Broadband/Dual-Frequency Two-Strip Antenna

To cover the DCS-1800 (DCS 1710–1880 MHz), PCS-1900 (1,895–1,918 MHz), PCS-1900 (1,850–1,990 MHz), IMT-2000/UMTS (1,885–2,200 MHz) for mobile communications, and the 5-GHz UNII band for wireless LANs (5.15–5.875 GHz), a dual-frequency antenna with a bandwidth of ~30% at the 2-GHz band and a bandwidth of ~15% at the 5-GHz band is required. In this section, we combine a two-strip monopole and a planar monopole onto a very thin substrate to achieve a broadband dual-frequency operation. The configuration of the dual-frequency planar antenna is illustrated in Figure 3.54 [10]. The dual-frequency antenna consists of a two-strip monopole for the 2-GHz band and a planar monopole for the 5-GHz band. The two-strip monopole is formed by an S-strip and a T-strip. The planar monopole and the T-strip are printed on the front side of the Duroid substrate and fed by a 50Ω microstrip line while the S-strip is etched on the backside of the substrate and terminated at a ground plane (its length=L_g). The upper section (its width=W_T and its height=H_T) of the T-strip is fitted (leaving a space of w_s) into an area surrounded by the upper section (its width=W_S) of the S-strip (its strip width=w_s) while the lower section of the T-strip overlaps with the lower section of the S-strip, forming a two-strip line. The width of the two-strip line is equal to the width (w_f) of the 50Ω microstrip line. The total height of the dual-frequency antenna is equal to the height (H) of the two-strip monopole. The planar monopole is equally divided into a wider upper part (its width=W_U and its height=H/2) and a narrower lower part (its width=W_L and its height=H/2) that is connected to the 50Ω microstrip line through a narrower microstrip line (its width=w_c and its length is equal to the total width (W_p) of the planar monopole). The total width (W) of the dual-fre-

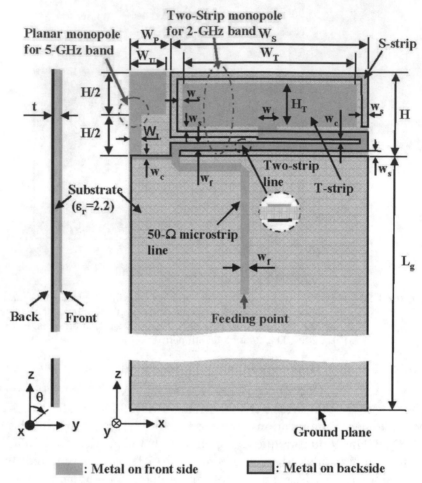

Figure 3.54 Geometry of the dual-frequency planar antenna that consists of a two-strip monopole for the 2-GHz band and a planar monopole for the 5 GHz band. (From: [10]. ©2008 IEEE. Reprinted with permission.)

quency antenna is equal to the sum of the widths of the two-strip monopole (W_S) and the planar monopole (W_P) (i.e., $W=W_S+W_P$). There is no direct electrical connection (e.g., by a shorting via) between the front side and the backside.

We began the design of the dual-frequency antenna with the two-strip monopole that was adjusted to have good impedance matching around 2 GHz. Then, a planar monopole was added to the two-strip monopole to achieve impedance matching around 5 GHz. Figure 3.55 shows the results for return loss of the two-strip monopole and the planar monopole when they operate alone (see the inset of the figure). Good impedance matching is observed at 2 GHz for the two-strip monopole and at 5 GHz for the planar monopole. Note that the bandwidths at the 2-GHz band and at the 5-GHz band are not satisfactory and require further enhancement.

To realize the dual-frequency operation, we feed the two-strip monopole and the planar monopole simultaneously with the same 50Ω microstrip line. To make the dual-frequency antenna have good impedance matching and sufficient bandwidths at the 2- and 5-GHz bands, the geometric parameters of the antenna, such

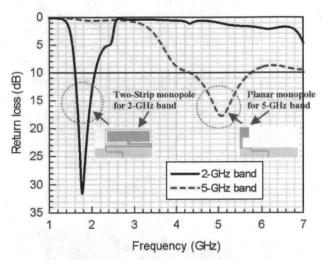

Figure 3.55 Comparison of the return loss between the two-strip monopole for the 2-GHz band and the planar monopole for the 5-GHz band (H=8 mm, W_S=19 mm, H_T=4.25 mm, W_T=17 mm, W_P=4 mm, W_U=3.5 mm, W_L=1 mm, w_s=0.5 mm, w_t=1.5 mm, w_c=0.25 mm, w_f=0.75 mm, t=0.254 mm, and L_g=60 mm). (From: [10]. ©2008 IEEE. Reprinted with permission.)

as H, W_S, H_T, W_T, W_P, W_U, W_L, w_s, and w_c, need to be optimized. Figure 3.56 shows the final result of return loss for an optimal dual-frequency planar antenna. It is found that the bandwidths (for return loss <-10 dB) at the 2- and 5-GHz bands are 35% (1.58–2.27 GHz) and 15% (4.98–5.75 GHz), respectively. The total height and width of the antenna are H=8.0 mm and W=23 mm, respectively. To verify the performance of the dual-frequency antennas, two prototypes were fabricated and measured: one with a height of 8.0 mm and the other with a height

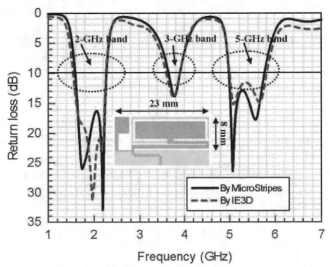

Figure 3.56 Return loss of the dual-frequency planar antenna simulated by microstrips and IE3D. (H=8 mm, W_S=19 mm, H_T=4.25 mm, W_T=17 mm, W_P=4 mm, W_U=3.5 mm, W_L=1 mm, w_s=0.5 mm, w_t=1.5 mm, w_c=0.25 mm, w_f=0.75 mm, t=0.254 mm, and L_g=60 mm). (From: [10]. ©2008 IEEE. Reprinted with permission.)

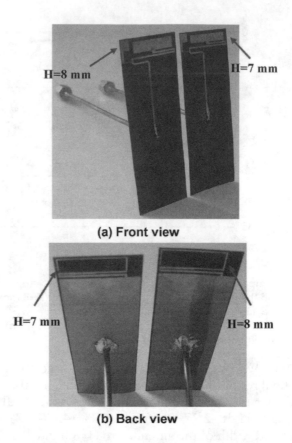

Figure 3.57 Photographs of two dual-frequency planar antennas that have antenna heights of 7.0 and 8.0 mm, respectively. (W_S=19 mm, W_T=17 mm, W_P=4 mm, W_U=3.5 mm, W_L=1 mm, w_s=0.5 mm, w_t=1.5 mm, w_c=0.25 mm, w_f=0.75 mm, t=0.254 mm, and L_g=60 mm). (a)Front view, and (b) back view. (From: [10]. ©2008 IEEE. Reprinted with permission.)

of 7.0 mm. Two photographs of the two prototypes are displayed in Figure 3.57. The measured results for return loss are presented in Figure 3.58. The measured bandwidth for the antenna with a height of 8.0 mm was found to be 32% (1.68–2.32 GHz) at the 2-GHz band and 15% (4.95–5.80 GHz) at the 5-GHz band. The frequency band for the antenna with a height of 7.0 mm was slightly shifted up to 1.70–2.38 GHz at the 2-GHz band and 5.05–5.95 GHz at the 5-GHz band, which entirely covers the 2-GHz band for mobile communications and the 5-GHz UNII band for WLANs.

The radiation pattern of the dual-frequency planar antenna is almost independent of the antenna height H. Also, the radiation pattern has no significant change over each frequency band.

3.9 Conclusions

In this chapter, we have introduced planar monopole antennas, inverted-L antennas, invert-F antennas, T-shaped antennas, meandered antennas, M-shaped antenna, and two-strip antennas for multiband operation. All of these multiband

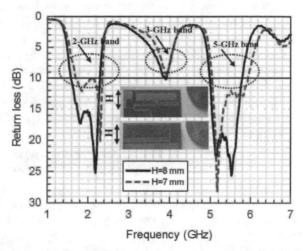

Figure 3.58 Measured return loss of two dual-frequency planar antennas with heights of 7.0 and 8.0 mm, respectively. (From: [10]. ©2008 IEEE. Reprinted with permission.)

antennas can be thought of as a derivative of a conventional monopole antenna, which have a fully planar configuration and may be integrated into a single PCB. For a multiband operation, multiple monopoles with different lengths can be combined and fed at a single point. To lower the antenna profile, the monopoles can be folded into different forms, such an L shape, F shape, T shape, S shape, meandered shape, or an M shape. The bandwidth of a multiband antenna can be enhanced by introducing a coupling feed or by adding some parasitic elements. It has been demonstrated that the multiband antennas introduced in this chapter can cover the frequency bands for 1G, 2G, 3G, or even 4G wireless applications.

References

[1] Balanis, C. A., *Antenna Theory: Analysis and Design*, New York: John Wiley & Sons, 2005, pp. 151–230.

[2] Chen, H.-M., et al., "A Compact Dual-Band Microstrip-Fed Monopole Antenna," *Proc. IEEE Antennas and Propagation Society Int. Symposium,* July 2001, Vol. 2, pp. 124–127.

[3] Nakano, H., et al., "An Inverted FL Antenna for Dual-Frequency Operation," *IEEE Transactions on Antennas and Propagation*, Vol. 53, No. 8, August 2005, pp. 2417–2421.

[4] Chen, S.-B., et al., "Modified T-Shaped Planar Monopole Antennas for Multiband Operation," *IEEE Transactions on Microwave Theory and Techniques*, Vol. 54, No. 8, August 2006, pp. 3267–3270.

[5] Wang, Y.-S., M.-C. Lee, and S.-J. Chung, "Two PIFA-Related Miniaturized Dual-Band Antennas," *IEEE Transactions on Antennas and Propagation*, Vol. 55, No. 3, March 2007, pp. 805–811.

[6] Liu, W.-C., W.-R. Chen, and C.-M. Wu, "Printed Double S-Shaped Monopole Antenna for Wideband and Multi-Band Operation of Wireless Communications," *IEE Proceedings on Microwaves, Antennas and Propagation*, Vol. 151, No. 6, December 2004, pp. 473–476.

[7]   Wang, H., "Dual-Resonance Monopole Antenna with Tuning Stubs," *IEE Proceedings on Microwaves, Antennas and Propagation*, Vol. 153, No. 4, August 2006, pp. 395–399.

[8]   Du, Z., K. Gong, and J. S. Fu, "A Novel Compact Wide-Band Planar Antenna for Mobile Handsets," *IEEE Transactions on Antennas and Propagation*, Vol. 54, No. 2, February 2006, pp. 613–619.

[9]   Li, R. L., et al., "A Compact Broadband Planar Antenna for GPS, DCS-1800, IMT-2000, and WLAN Applications," *IEEE Antennas and Wireless Propagation Letters*, Vol. 6, 2007, pp. 25–27.

[10]   Li, R. L., et al., "A Novel Low-Profile Dual-Frequency Planar Antenna for Wireless Handsets," *IEEE Transactions on Antennas and Propagation*, Vol. 56, No. 4, April 2008.

[11]   Chen, H.-D., and H.-T. Chen, "A CPW-Fed Dual-Frequency Monopole Antenna," *IEEE Transactions on Antennas and Propagation*, Vol. 52, No. 4, April 2004, pp. 978–982.

[12]   Liu, W.-C., "Design of a Multi-Band CPW-Fed Monopole Antenna Using a Particle Swarm Optimization Approach," *IEEE Transactions on Antennas and Propagation*, Vol. 53, No. 10, October 2005, pp. 3273–3279.

[13]   John, M., and M. J. Ammann, "Integrated Antenna for Multi-Band Multi-National Wireless Combined with GSM1800/PCS1900/IMT2000 + Extension," *Microwave and Optical Technology Letters*, Vol. 48, No. 9, March 2006, pp. 613–615.

[14]   Wong, K.-L., G.-Y. Lee, and T.-W. Chiou, "A Low-Profile Planar Monopole Antenna for Multi-Band Operation of Mobile Handsets," *IEEE Transactions on Antennas and Propagation*, Vol. 51, No. 1, January 2003, pp. 121–125.

[15]   Jan, J.-Y., and L.-C. Tseng, "Small Planar Monopole Antenna with a Shorted Parasitic Inverted-L Wire for Wireless Communications in the 2.4-, 5.2- and 5.8-GHz Bands," *IEEE Transactions on Antennas and Propagation*, Vol. 52, No. 7, July 2004, pp. 1903–1905.

[16]   Kuo, Y.-L., and K.-L. Wong, "Printed Double-T Monopole Antenna for 2.4/5.2 GHz Dual-Band WLAN Operations," *IEEE Transactions on Antennas and Propagation*, Vol. 51, No. 9, September 2003, pp. 2187–2192.

[17]   Wong, K.-L. L.-C. Chou, and C.-M. Su, "Dual-Band Flat-Plate Antenna with Shorted Parasitic Element for Laptop Applications," *IEEE Transactions on Antennas and Propagation*, Vol. 53, No. 1, January 2005, pp. 539–544.

[18]   Wu, C.-M., C.-N. Chiu, and C.-K. Hsu, "A New Nonuniform Meandered and Fork-Type Grounded Antenna for Triple-Band WLAN Applications," *IEEE Antennas and Wireless Propagation Letters*, Vol. 5, 2006, pp. 346–348.

[19]   Jing, X., Z. Du, and K. Gong, "A Compact Multi-Band Planar Antenna for Mobile Handsets," *IEEE Antennas and Wireless Propagation Letters*, Vol. 5, 2006, pp. 343–345.

CHAPTER 4

Printed Multiband Fractal Antennas

Wojciech J. Krzysztofik

4.1 Introduction

4.1.1 Brief Background on Fractal Geometry

Recent efforts by several researchers [1] around the world to combine fractal geometry with electromagnetic theory have led to a plethora of new and innovative antenna designs. In this chapter, we provide a comprehensive overview of recent developments in the rapidly growing field of fractal antenna engineering, which has been primarily focused in two areas: the analysis and design of fractal antenna elements, and the application of fractal concepts to the design of antenna arrays.

It has been an intriguing question among the electromagnetic community [2] as to what property of fractals, if any, is really useful, especially when it comes to designing fractal shaped antenna elements. Several fractal geometries have been explored for antennas with special characteristics, in the context of both antenna elements and spatial distribution functions for elements in antenna arrays. Yet there have been no concerted efforts to design antennas using fractal shaped elements. In many fractal antennas, the self-similarity and plane-filling nature of fractal geometries are often qualitatively linked to its frequency characteristics.

The fractal geometry has been behind an enormous change in the way scientists and engineers perceive, and subsequently model, the world in which we live [3]. Chemists, biologists, physicists, physiologists, geologists, economists, and engineers (mechanical, electrical, chemical, civil, aeronautical) have all used methods developed in fractal geometry to explain a multitude of diverse physical phenomena: from trees to turbulence, cities to cracks, music to moon craters, measles epidemics, and much more. Many of the ideas within fractal geometry have been in existence for a long time; however, it took the arrival of the computer, with its capacity to accurately and quickly carry out large repetitive calculations, to provide the tool necessary for the in-depth exploration of these subject areas. In recent years, the explosion of interest in fractals has essentially ridden on the back of advances in computer development.

4.1.1.1 A Matter of Fractals

In recent years, the science of fractal geometry has grown into a vast area of knowledge, with almost all branches of science and engineering gaining from the new insights it has provided. Fractal geometry is concerned with the properties of fractal objects, usually simply known as fractals (see Figure 4.1).

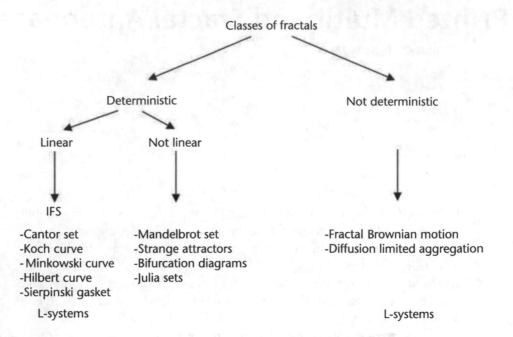

Figure 4.1   Classes of fractals.

Fractals may be found in nature or generated using a mathematical recipe. The word "fractal" was coined by Benoit Mandelbrot, sometimes referred to as the father of fractal geometry, who said, "I coined *fractal* from the Latin adjective *fractus*. The corresponding Latin verb *frangere* means 'to break' to create irregular fragments. It is therefore sensible—and how appropriate for our need!—that, in addition to 'fragmented' (as in *fraction* or *refraction*), *fractus* should also mean 'irregular', both meanings being preserved in fragment" [4]. Moreover he asked: "Why geometry is often described as 'cold' or 'dry'? One reason lies in its inability to describe the shape of a cloud, a mountain, a coastline, or a tree. Clouds are not spheres, mountains are not cones, coastlines are not circles, and bark is not smooth, nor does lightning travel in a straight line."

To date, there exists no watertight definition of a fractal object. Mandelbrot offered the following definition: "A fractal is by definition a set for which the Hausdorff dimension strictly exceeds the topological dimension," which he later retracted and replaced with: "A fractal is a shape made of parts similar to the whole in some way."

So, possibly the simplest way to define a fractal is as an object that appears self-similar under varying degrees of magnification, and in effect, possessing symmetry across scale, with each small part of the object replicating the structure of the whole. This is perhaps the loosest of definitions; however, it captures the essential, defining characteristic, that of self-similarity.

But here are five properties that most fractals have:

1. Fractals have details on arbitrarily small scales.
2. Fractals are usually defined by simple recursive processes.

3. Fractals are too irregular to be described in traditional geometric language.
4. Fractals have some sort of self-similarity.
5. Fractals have fractal dimension.

The general comparison of Euclidean and fractal geometries is shown in Table 4.1.

4.1.1.2   Random Fractals

The exact structure of regular fractals is repeated within each small fraction of the whole (i.e., they are exactly self-similar). There is, however, another group of fractals, known as random fractals, which contain a random or statistical elements. These fractals are not exactly self-similar, but rather statistically self-similar. Each small part of a random fractal has the same statistical properties as the whole. Random fractals are particularly useful in describing the properties of many natural objects and processes. A simple way to generate a fractal with an element of randomness is to add some probabilistic element to the construction process of a regular fractal. While such random fractals do not have the self-similarity of their nonrandom counterparts, their nonuniform appearance is often rather closer to natural phenomena such as coastlines, topographical surfaces, or cloud boundaries. Indeed, random fractal constructions are the basis of many impressive computer-drawn landscapes or skyscapes. A random fractal worthy of the name should display randomness at all scales, so it is appropriate to introduce a random element at each stage of the construction. By relating the size of the random variations to the scale, we can arrange for the fractal to be statistically self-similar in the sense that enlargements of small parts have the same statistical distribution as the whole set. This compares with (nonrandom) self-similar sets where enlargements of small parts are identical to the whole.

4.1.2   Fractals in Nature

The original inspiration for the development of fractal geometry came largely from an in-depth study of the patterns of nature. For instance, fractals have

Table 4.1   General Comparison of Euclidean and Fractal Geometries

Euclidean Versus Fractal Geometry	
Euclidean Geometry	*Fractal Geometry*
> 2,000 years old	10–20 years old
Applicable for artificial objects	Applicable for natural objects
Shapes change with scaling	Invariant under scaling, self-similar
Objects defined by analytical equations	Objects defined by recursive algorithms
Locally smooth, differentiable	Locally rough, not differentiable
Elements: vertices, edges, surfaces	Elements: iteration of functions

been successfully used to model such complex natural objects as galaxies, cloud boundaries, mountain ranges, coastlines, snowflakes, trees, leaves, ferns, and much more.

Mandelbrot realized that it is very often impossible to describe nature using only Euclidean geometry that is in terms of straight lines, circles, cubes, and so forth. He proposed that fractals and fractal geometry could be used to describe real objects, such as trees, lightning, river meanders, and coastlines, to name but a few. Figure 4.2 contains some natural fractals: the boundary of clouds, a hillside silhouette, plants, the human lungs, and fractal arts [5].

Fractal geometries have found an intricate place in science as a representation of some of unique geometrical features occurring in nature. Fractals are used to describe the branching of tree leaves and plants, the space filling of water vapor

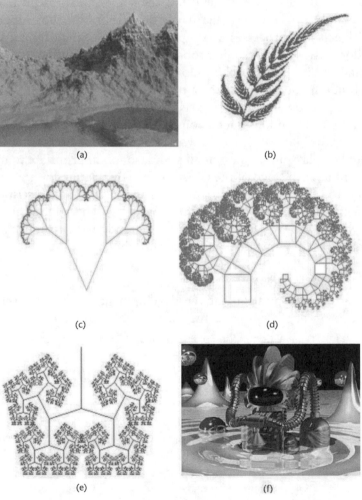

Figure 4.2   Natural fractal objects: landscape by Roger Barton, where clouds, coastlines, trees, and water surface were created using fractals (a), plants a tree branch looks similar to the entire tree, a fern leaf looks almost identical to the entire fern, one classic way of creating fractal plants is by means of L-systems (Lindenmayer) (b), (c), (d) the human lungs (e) and the fractal art (f) [5].

that forms clouds, the random erosion that carves mountain faces, the jaggedness of coastlines and bark, and many more examples in nature. The structure of our universe—superclusters, clusters, galaxies, star systems (solar system), planets, moons—every detail of the universe shows the same clustering patterns. It can be modeled by random Cantor fractal square.

Originally, it was believed that Saturn had only a single ring. After some time, a break in the middle was discovered, and scientists concluded that there were two rings. However, when Voyager I approached Saturn, it was discovered that the two rings were also broken in the middle, and these four smaller rings were broken as well. Eventually, Voyager I identified a very large number of breaks, which continuously broke even small rings into smaller pieces. The overall structure is amazingly similar to the Cantor fractal set.

Weather behaves very unpredictably; sometimes it changes very smoothly, and other times it changes very rapidly. Edward Lorenz came up with three formulas that could model the changes of the weather. These formulas are used to create a 3-D strange attractor; they form the famous Lorenz Attractor, which is a fractal pattern.

In the human body the lungs, formed by splitting lines, are fractal canopies; similarly, the surface of the brain contains a large number of folds that are well modeled by fractal shapes.

During electronic transmissions, electronic noise would sometimes interfere with the transmitted data. Although making the signal more powerful would drown out some of this harmful noise, some of it persisted, creating errors during transmissions. Errors occurred in clusters: a period of no errors would be followed by a period with many errors. On any scale of magnification (month, day, hour, 20 minutes), the proportion of error-free transmission to error-ridden transmission stays constant. Mandelbrot studied the mathematical process that enables us to create random Cantor dust, describing perfectly well the fractal structure of the batches of errors on computer lines.

Many more examples could be introduced to prove the fractal nature of universe. Therefore, there is a need for a geometry that handles these complex situations better than Euclidean geometry.

4.1.3   Prefractals: Truncating a Fractal to Useable Complexity

There is some terminology that should be established to understand fractals and how they can be applied to practical applications [6]. Even though a fractal is mathematically defined to be infinite in intricacy, this is obviously not the case in nature, and is not desirable if some objects in nature are to be modeled using these geometries. For example, the complexity and repetition of a cloud does not extend to infinitely small or large scales, but can be approximated as doing so for a certain band of scales. From the scale of human perception, a cloud does seem to be infinitely complex in larger and smaller scales. The resulting geometry after truncating the complexity is called a prefractal [7]. A prefractal drops the intricacies that are not distinguishable in the particular applications. In Figure 4.3, the fifth iteration is indistinguishable from the Cantor set obtained at higher iterations.

This problem occurs due to the limit of the finite detail our eyes (or the printer we use to plot the image) can resolve. Thus, to illustrate the set, it is sufficient to repeat the generation process only by the number of steps necessary to fool the

Figure 4.3   A uniform Cantor fractal set.

eye, and not an infinite number of times. This is true for all illustrations of fractal objects. However, make no mistake, only after an infinite number of iterations do we obtain the Cantor set. For a finite number of iterations the object produced is merely a collection of line segments with finite measurable length. These objects formed en route to the fractal object are termed prefractals.

4.1.4   How Fractals Can Be Used as Antennas and Why Fractals Are Space-Filling Geometries

While Euclidean geometries are limited to points, lines, sheets, and volumes, fractals include the geometries that fall between these distinctions. Therefore, a fractal can be a line that approaches a sheet. The line can meander in such a way as to effectively almost fill the entire sheet. These space-filling properties lead to curves that are electrically very long, but fit into a compact physical space. This property can lead to the miniaturization of antenna elements. In the previous section, it was mentioned that prefractals drop the complexity in the geometry of a fractal that is not distinguishable for a particular application. For antennas, this can mean that the intricacies that are much, much smaller than a wavelength in the band of useable frequencies can be dropped out [8]. This now makes this infinitely complex structure, which could only be analyzed mathematically, but may not be possible to be manufactured. It will be shown that the band of generating iterations required to reap the benefits of miniaturization is only a few before the additional complexities become indistinguishable. There have been many interesting works that have looked at this emerging field of fractal electrodynamics. Much of the pioneering work in this area has been documented in [9, 10]. These works include fundamentals about the mathematics as well as studies in fractal antennas and reflections from fractal surfaces.

The space-filling properties of the Hilbert curve and related curves (e.g., Peano fractal) make them attractive candidates for use in the design of fractal antennas. The space-filling properties of the Hilbert curve were investigated in [11, 12] as an effective method for designing compact resonant antennas. The first four steps in the construction of the Hilbert curve are shown in Figure 4.4.

The Hilbert curve is an example of a space-filling fractal curve that is self-voiding (i.e., has no intersection points). In the antenna engineering it can be used as dipole (fed in the center), monopole (fed on one side) antenna, as well as meandered structure of microstrip patch antenna.

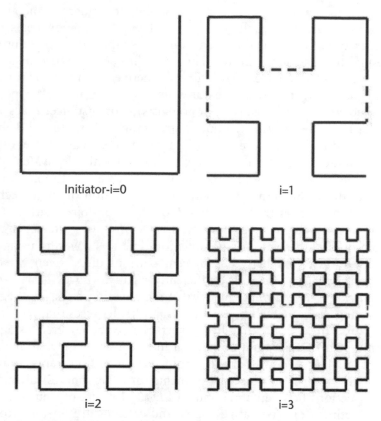

Initiator-i=0   i=1

i=2   i=3

Figure 4.4   Generation of four iterations of Hilbert curves. The segments used to connect the geometry of the previous iteration are shown in dashed lines.

4.1.5 The Fractal Dimensions of Box Counting, Hausdorff, and Similarity

Generally, we can conceive of objects that are zero dimensional (0-D) (points), 1-D (lines), 2-D (planes), and 3-D (solids). We feel comfortable with zero, one, two, and three dimensions. We form a 3-D picture of our world by combining the 2-D images from each of our eyes. Is it possible to comprehend higher-dimensional objects (i.e., 4-D, 5-D, 6-D)? What about noninteger-dimensional objects such as 2.12-D, 3.79-D or 36.91232 . . . -D? There are many definitions of fractal dimension and we shall encounter a number of them as we proceed through the text, including the similarity dimension (DS), the divider dimension (DD), the Hausdorff dimension (DH), the box-counting dimension (DB), the correlation dimension (DC) the information dimension (DI), the pointwise dimension (DP), the averaged pointwise dimension (DA), and the Lyapunov dimension (DL). The last seven dimensions listed are particularly useful in characterizing the fractal structure of strange attractors associated with chaotic dynamics.

So, the main tool for fractal geometry is the dimension in it many forms. Before we deal with fractal dimensions, let us look at the two most common, and perhaps most comprehensible, definitions of dimension, the Euclidean dimension (DE), and topological dimension (DT) [3]. Both definitions lead to nonfractal, integer dimensions. The Euclidean dimension is simply the number of coordinates required to specify the object. The topological dimension is more involved. The branch of mathematics known as topology considers shape and form of objects from essentially a qualitative point of view. Topology deals with the ways in which objects may be distorted from one shape and formed into another without losing their essential features. Thus straight lines may be transformed into smooth curves or bent into "crinkly" curves, where each of the constructions is topologically equivalent. Certain features are invariant under proper transformations (called homeomorphisms by topologists)—for instance, holes in objects remain holes regardless of the amount of stretching and twisting the object undergoes in its transformation from one shape to another. All of the two-holed surfaces, although quite different in shape, are topologically equivalently as each one may be stretched and molded into one of the others. The topological dimension of an object does not change under the transformation of the object. The topological dimension derives from the ability to cover the object with discs of small radius. The line segment may be covered using many discs intersecting many times with each other. However, it is possible to refine this covering using discs with only a single intersection between adjacent pairs of discs. Even when the line is contorted, one can find discs sufficiently small to cover it with only intersections occurring between adjacent pairs of the covering discs. The segment within each covering disc can itself be covered using smaller discs that require only to intersect in pairs. In a similar manner, a surface may be covered using spheres of small radius with a minimum number of intersections requiring intersecting triplets of spheres. The definition of the topological dimension stems from this observation. The covering of an object by elements (discs or spheres) of small radius requires intersections between a minimum of $DT + 1$ group of elements.

Methods of classical geometry and calculus are unsuited to studying fractals and we need alternative techniques [7]. For example, Hausdorff dimension (DH) and the box-counting dimension (DB) may be defined for any sets, and may be shown to equal to the similarity dimension (DS). Very roughly, a dimension provides a description of how much space a set fills. It is a measure of the prominence of irregularities of a set when viewed at very small scales. A dimension contains much information about the geometrical properties of a set.

One of the most important in classifying fractals is the Hausdorff dimension [3]. In fact, Mandelbrot suggested that a fractal may be defined as an object that has a Hausdorff dimension which exceeds its topological dimension. A complete mathematical description of the Hausdorff dimension is outside the scope of this text. In addition, the Hausdorff dimension is not particularly useful to the engineer or scientist hoping to quantify a fractal object, the problem being that it is practically impossible to calculate it for real data. We therefore begin this section by concentrating on the closely related box-counting dimension and its application to determining the fractal dimension of natural fractals before coming to a brief explanation of the Hausdorff dimension.

4.1.5.1 The Box-Counting Dimension

The box-counting or box dimension (DB) is one of the most widely used dimensions [3, 7]. Its popularity is largely due to its relative ease of mathematical calculation and empirical estimation. The definition goes back at least to the 1930s and it has been variously termed Kolmogorov entropy, entropy dimension, and metric dimension. In this text, we will always refer to box or box-counting dimension to avoid confusion.

Let F be any nonempty bounded subset of $\Re^n$ (in n-dimensional Euclidian space $\Re^n$) and let $N_\delta(F)$ be the smallest number of sets of diameter at most δ that can cover F. The box-counting dimension or box dimension of F is defined as

$$DB(F) = \lim_{\delta \to 0} \frac{\log N_\delta(F)}{-\log \delta} \qquad (4.1)$$

There are several equivalent definitions of box dimension that are sometimes more convenient to use.

To examine a suspected fractal object for its box-counting dimension we cover the object with covering elements or "boxes" of side length δ [3]. The number of boxes, N, required to cover the object is related to δ through its box-counting dimension, DB. The method for determining DB is illustrated in the simple example, where a straight line (a one-dimensional object) of unit length is probed by cubes (3-D objects) of side length δ. We require N cubes (volume δ^3) to cover the line. Similarly, if we had used squares of side length δ (area δ^2) or line segments (length δ^1), we would again have required N of them to cover the line. Equally, we could also have used 4-D, 5-D, or 6-D elements to cover the line segment and still required just N of them. In fact, to cover the unit line segment, we may use any elements with dimension greater than or equal to the dimension of the line itself, namely one. To simplify matters, the line is specified as exactly one unit in length. The number of cubes, squares, or line segments we require to cover this unit line is then $N\delta$ (=1), hence $N=1/\delta^1$. Notice that the exponent of δ remains equal to one regardless of the dimension of the probing elements, and is in fact the box-counting dimension, DB, of the object under investigation. Notice also that for the unit (straight) line, the Euclidean, the box, and the topological-dimension, are equal each other, $DE=DB=DT$ (= 1), hence it is not a fractal by our definition in Section 4.1.1, as the fractal dimension, here given by DB, does not exceed the topological dimension, DT.

If we repeat the covering procedure outlined above for a plane unit area, it is easy to see that to cover such a unit area we would require $N=1/\delta^2$ hypercubes of edge length δ and Euclidean dimension greater than or equal to two. Similarly, with a 3-D solid object we would require $N=1/\delta^3$ hypercubes of edge length δ with Euclidean dimension greater than or equal to three to cover it. Again notice that in each case the exponent of is a measure of the dimension of the object. In general, we require $N = 1/\delta^{DB}$ boxes to cover an object where the exponent DB is the box-counting dimension of the object. We arrive at the following general formulation of DB for objects of unit hypervolume:

$$DB = \frac{\log(N)}{\log(1/\delta)} \qquad (4.2)$$

obtained by covering the object with N hypercubes of side length δ. Note that the above expression is of rather limited use. It assumes the object is of unit hypervolume and in general will produce erroneous results for large δ. More general and practically useful expressions are given in (4.3). Note also the marked resemblance of (4.2) to the definition of the similarity dimension DS given in (4.1).

However, do not confuse the two: the calculation of DS requires that exactly self-similar parts of the fractal are identified, whereas DB requires the object to be covered with self-similar boxes. Hence, DB allows us greater flexibility in the type of fractal object that may be investigated.

The general expression for the dimension of an object with a hypervolume (i.e., length, area, volume, or fractal hypervolume) not equal to unity, but rather given by V^*, is

$$DB = \frac{\log(N) - \log(V^*)}{\log(1/\delta)} \tag{4.3a}$$

where N is the number of hypercubes of side length δ required to cover the object (i.e., $N = V^*/\delta^{DB}$). Rearranging (4.3a) gives

$$\log(N) = DB \log(1/\delta) + \log(V^*) \tag{4.3b}$$

which is in the form of the equation of a straight line where the gradient of the line, DB, is the box-counting dimension of the object. This form is suitable for determining the box-counting dimension of a wide variety of fractal objects by plotting $\log(N)$ against $\log(1/\delta)$ for probing elements of various side lengths, δ. The box-counting dimension is widely used in practice for estimating the dimension of a variety of fractal objects. The technique is not confined to estimating the dimensions of objects in the plane, such as the coastline curve. It may be extended to probe fractal objects of high fractal dimension in multidimensional spaces, using multidimensional-covering hypercubes. Its popularity stems from the relative ease by which it may be incorporated into computer algorithms for numerical investigations of fractal data. The grid method lends itself particularly to encoding within a computer program. By covering the data with grids of different box side lengths, δ, and counting the number of boxes, N, that contain the data, the box-counting dimension is easily computed using (4.3b).

4.1.5.2 Hausdorff Measure and Dimension

Of the wide variety of fractal dimensions in use, the definition of Hausdorff, based on a construction of Carathéodory, is the oldest and probably the most important [7]. Hausdorff dimension has the advantage of being defined for any set, and is mathematically convenient, as it is based on measures, which are relatively easy to manipulate. A major disadvantage is that in many cases it is hard to calculate or to estimate by computational methods. However, for an understanding of the mathematics of fractals, familiarity with Hausdorff measure and dimension is essential.

We generally work in n-dimensional Euclidean space $\mathfrak{R}^n$, where $\mathfrak{R}^1=\mathfrak{R}$ is just the set of real numbers or the "real line," $\mathfrak{R}^2$ is the (Euclidean) plane. Points in $\mathfrak{R}^n$ will generally be denoted by lowercase letters x, y, and so forth.

Sets, which will generally be subset of $\mathfrak{R}^n$, are denoted by capital letters E, F, U, and so forth. In the usual way, $x \in E$ means that the point x belongs to set E, and $E \subset F$ means that E is subset of the set F. We write $\{x: \text{condition}\}$ for the set of x for which "condition" is true. Certain frequently occurring sets have special notation.

Recall that if U is any nonempty subset of n-dimensional Euclidean space, $\mathfrak{R}^n$.

The diameter of U is defined as $|U| = \sup\{|x - y| : x, y \in U\}$ (i.e., the greatest distance apart of any pair of points in U.) If $\{U_i\}$ is a countable (or finite) collection of sets of diameter at most δ that cover F (i.e., $F \subset \bigcup_{i=1}^{\infty} U_i$ with $0 < |U_i| \le \delta$ for each i), we say that $\{U_i\}$ is a δ-cover of F.

Suppose that F is a subset of $\mathfrak{R}^n$ and s is a nonnegative number. For any $\delta > 0$ we define

$$\mathrm{H}_{\delta}^s(F) = \inf\{\sum_{i=1}^{\infty} |U_i|^s : \{U_i\} \text{ is a } \delta - \mathrm{cov}er \text{ of } F\} \tag{4.4}$$

Thus, we look at all covers of F by sets of diameter at most δ and seek to minimize the sum of the sth powers of diameters. As δ decreases, the class of permissible covers of F is reduced. Therefore, the infimum $H^s{}_\delta(F)$ increases, and so approaches a limit as $\delta \to 0$. We write

$$\mathrm{H}^s(F) = \lim_{\delta \to 0} \mathrm{H}_{\delta}^s(F) \tag{4.5}$$

This limit exists for any subset F of $\mathfrak{R}^n$ though the limiting value can be (and usually is) 0 or ∞. We call the $H^s{}_\delta(F)$ the s-dimensional Hausdorff measure of F. With a certain amount of effort H^s may be shown to be a measure. In particular $H^s(\varnothing) = 0$, if E is contained in F then $H^s(E) \le H^s(F)$, and if $\{F_i\}$ is any countable collection of disjoint Borel set, then

$$H^s\left(\bigcup_{i=1}^{\infty} F_i\right) = \sum_{i=1}^{\infty} H^s(F_i) \tag{4.6}$$

Hausdorff measures generalize the familiar ideas of length, area, volume, and so forth. It may be shown that, for subsets of $\mathfrak{R}^n$, n-dimensional Hausdorff measure is, to within a constant multiple, just n-dimensional Lebesgue measure (i.e., the usual n-dimensional volume). More precisely, if F is a Borel subset of $\mathfrak{R}^n$, then

$$H^n(F) = c_n vol^n(F) \tag{4.7}$$

where the constant $c_n = \pi^{\frac{1}{2}n} / 2^n (\frac{1}{2}n)!$ is the volume of an n-dimensional ball of diameter 1.

Similarly, for "nice" lower-dimensional subsets of $\mathfrak{R}^n$, we have that $H^0(F)$ is the number of points in F; $H^1(F)$ gives the length of smooth curve F,

$$H^2(F) = \frac{1}{4}\pi \times area(F)$$

if F is a smooth surface,

$$H^3(F) = \frac{4}{3}\pi \times vol(F)$$

and

$$H^m(F) = c_m \times vol^m(F)$$

if F is a smooth m-dimensional submanifold of $\mathfrak{R}^n$ (i.e., an m-dimensional surface in classical sense).

The scaling properties of length, area, and volume are well known. On magnification by a factor λ, the length of a curve is multiplied by λ, the area of plane region is multiplied by λ^2 and the volume of a three-dimensional object is multiplied by λ^3. As might be anticipated, s-dimensional Hausdorff measure scales with a factor λ^s. Such scaling properties are fundamental to the theory of fractals.

Returning to (4.4) it is clear that for any given set F and $\delta<1$, $H^s_\delta(F)$ is nonincreasing with s, so $H^s_\delta(F)$ is also nonincreasing. In fact, rather more is true if $t > s$ and $\{U_i\}$ is a δ-cover of F we have

$$\sum_i |U_i|^t \le \delta^{t-s} \sum_i |U_i|^s \qquad (4.8)$$

so, taking infima, $H^t_\delta(F) \le \delta^{t-s} H^s_\delta(F)$. Letting $\delta \rightarrow 0$ we see that if $H^s(F) < \infty$ then $H^t(F)<0$ for $t>s$. It shows that there is a critical value of s at which $H^s(F)$ jumps' from ∞ to 0. This critical value is called the Hausdorff dimension of F, and written $DH(F)$. (Note that some authors refer to Hausdorff dimension as Hausdorff-Besicovitch dimension.) Formally

$$DH(F) = \inf\{s : H^s(F) = 0\} = \sup\{s : H^s(F) = \infty\} \qquad (4.9)$$

so that

$$H^s(F) = \begin{cases} \infty & if \quad s < DH(F) \\ 0 & if \quad s > DH(F) \end{cases} \qquad (4.10)$$

If $s=DH(F)$, then $H^s(F)$ may be zero or infinite, or may satisfy

$$0 < H^s(F) < \infty \qquad (4.11)$$

A Borel set satisfying this last condition is called an *s*-set. Mathematically *s*-sets are by far the most convenient sets to study, and fortunately this occurs surprisingly often.

For a very simple example, let *F* be a flat disc of unit radius in $\Re^n$. From familiar properties of length, area, and volume,

$$H^1(F) = length(F) = \infty \qquad (4.12)$$

$$0 < H^2(F) = 0.25\pi \times area(F) < \infty \text{ and } H^3(F) = \frac{4}{3}\pi \times vol(F) = 0 \quad (4.13)$$

Thus DH(F)=2 with $H^s(F) = \infty$ if $s < 2$ and $H^s(F) = 0$ if $s > 2$ (4.14)

It may be recalled that fractal dimension is an important characteristic of fractal geometries. However this is not a unique description for the geometry, but rather identifies a group of geometries with similar nature. Hence a first step in the utilization of fractal properties in antenna design should involve the dimension of the geometry [2]. We have found in cases where the appearance of the geometry is perturbed in such a way that its similarity dimension is varied, the multiband characteristics are affected, where as other perturbations leave this rather unaffected. This gives a further proof that antenna characteristics can in fact be linked to the fractal dimension of such geometries.

4.1.5.3 Similarity Dimension of Self-Similar Sets

Fractals are characterized by their dimension. It is the key structural parameter describing the fractal and is defined by partitioning the volume where the fractal lies into boxes of side δ. We hope that over a few decades in δ, the number of boxes that contain at least one of the discharge elements will scale as

$$N(\delta) \sim \delta^{-D} \qquad (4.15)$$

In the remaining parts of this chapter we will concentrate on the similarity dimension, denoted DS, to characterize the construction of regular fractal objects. As it was shown in previous chapters, different definitions of dimension are introduced where appropriate. The concept of dimension is closely associated with that of scaling. Consider the line, surface, and solid, divided up respectively by self-similar sublengths, subareas, and subvolumes of side length δ. For simplicity in the following derivation assume that the length, L, area, A, and volume, V, are all equal to unity. Consider first the line. If the line is divided into N smaller self-similar segments, each δ in length, then δ is in fact the scaling ratio (i.e., $\delta/L=\delta$, since $L=1$). Thus $L=N\delta=1$ (i.e., the unit line is composed of N self-similar parts scaled by $\delta=1/N$).

Now consider the unit area. If we divide the area again into N segments each δ^2 in area, then $A = N\delta^2 = 1$ (i.e., the unit surface is composed of N self-similar parts scaled by $\delta = 1/N^{1/2}$).

Applying similar logic, we obtain for a unit volume $V = n\delta^3 = 1$ (i.e., the unit solid is N self-similar parts scaled by $\delta = 1/N^{1/3}$).

Examining above expressions we see that the exponent of δ in each case is a measure of the (similarity) dimension of the object, and we have in general

$$N\delta^{DS} = 1 \qquad (4.16)$$

Using logarithms leads to the expression

$$DS = \frac{\log(N)}{\log(1/\delta)} \qquad (4.17)$$

Here the letter S denotes the similarity dimension.

The above expression has been derived using familiar objects that have the same integer Euclidean, topological, and similarity dimensions (i.e., a straight line, planar surface, and solid object), where $DE = DS = DT$. However, equation (4.17) may also be used to produce dimension estimates of fractal objects where DS is noninteger. This can be seen by applying the above definition of the self-similar dimension to the triadic Cantor set showed in Figure 4.3. We saw that the left-hand third of the set contains an identical copy of the set. There are two such identical copies of the set contained within the set, thus $N = 2$ and $\delta = 1/3$. According to (4.17) the similarity dimension is then

$$DS = \frac{\log(2)}{\log(1/(1/3))} = \frac{\log(2)}{\log(3)} = 0.6309...$$

Thus, for the Cantor set, DS is less than one and greater than zero: in fact it has a noninteger similarity dimension of $0.6309\ldots$ due to the fractal structure of the object. We saw in the previous section that the Cantor set has Euclidean dimension of one and a topological dimension of zero, thus $DE > DS > DT$. As the similarity dimension exceeds the topological dimension, the set is a fractal with a fractal dimension defined by the similarity dimension of $0.6309\ldots$ As an aid to comprehension it may be useful to think of the Cantor set as neither a line nor a point, but rather something in between.

Instead of considering each subinterval of the Cantor set scaled down by one-third we could have looked at each subinterval scaled down by one-ninth. As we saw from Figure 4.3, there are four such segments, each an identical copy of the set. In this case $N = 4$ and $\delta = 1/9$ and again this leads to a similarity dimension of

$$DS = \frac{\log(4)}{\log(1/(1/9))} = \frac{\log(4)}{\log(9)} = \frac{2\log(2)}{2\log(3)} = 0.6309...$$

4.2 Fractals Defined by Transformations—Self-Similar and Self-Affine Sets

4.2.1 Iterated Function Schemes: The Language of Fractals

Iterated function systems (IFS) represent an extremely versatile method for conveniently generating a wide variety of useful fractal structures [1, 3, 7, 13]. These iterated function systems are based on the application of a series of affine transformations, w, defined by [1]

$$w\begin{pmatrix} x \\ y \end{pmatrix} = \begin{pmatrix} a & b \\ c & d \end{pmatrix}\begin{pmatrix} x \\ y \end{pmatrix} + \begin{pmatrix} e \\ f \end{pmatrix} \tag{4.18}$$

or, equivalently, by

$$w(x, y) = (ax + by + e, \; cx + dy + f) \tag{4.19}$$

where a, b, c, d, e, and f are real numbers. Hence, the affine transformation, w, is represented by six parameters

$$\begin{pmatrix} a & b & e \\ c & d & f \end{pmatrix} \tag{4.20}$$

whereby a, b, c, and d control rotation and scaling, while e and f control linear translation (see Figure 4.5).

Now suppose we consider w_1, w_2, ..., w_N as a set of affine linear transformations, and let A be the initial geometry. Then a new geometry, produced by applying the set of transformations to the original geometry, A, and collecting the results from $w_1(A)$, $w_2(A)$, ..., $w_N(A)$, can be represented by

$$W(A) = \bigcup_{n=1}^{N} w_n(A) \tag{4.21}$$

where W is known as the Hutchinson operator [3, 13].

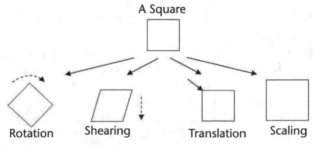

Figure 4.5 The affine transforms.

A fractal geometry can be obtained by repeatedly applying W to the previous geometry. For example, if the set A_0 represents the initial geometry, then we will have

$$A_1 = W(A_0); \quad A_2 = W(A_1); \quad \dots \quad ; \quad A_{k+1} = W(A_k) \qquad (4.22)$$

An iterated function system generates a sequence that converges to a final image, A_∞, in such a way that

$$W(A_\infty) = A_\infty \qquad (4.23)$$

This image is called the attractor of the iterated function system, and represents a "fixed point" of W.

For the Koch fractal curve the matrix of affine transformation has following form

$$w_q \begin{pmatrix} x \\ y \end{pmatrix} = \begin{pmatrix} \delta_{q1} \cos\theta_{q1} & -\delta_{q2} \sin\theta_{q2} \\ \delta_{q1} \sin\theta_{q1} & \delta_{q2} \cos\theta_{q2} \end{pmatrix} \begin{pmatrix} x \\ y \end{pmatrix} + \begin{pmatrix} t_{q1} \\ t_{q2} \end{pmatrix} \qquad (4.24)$$

and scaling factor is expressed as

$$\delta_q = \frac{1}{2 + 2\cos\theta_q} \qquad (4.25)$$

where θ_{qi} is the inclination angle of the second subsection with respect to the first, and t_{qi} is an element displacement on the plane. Figure 4.6 illustrates the iterated function system procedure for generating the well-known Koch fractal curve. In this case, the initial set, A_0, is the line interval of unit length (i.e., $A_0 = \{x: x \in [0,1]\}$, $\theta_q = 60°$, and $\delta_q = \frac{1}{3}$). Four affine linear transformations are then applied to A_0, as indicated in Figure 4.6. Next, the results of these four linear transformations are combined together to form the first iteration of the Koch curve, denoted by A_1. The second iteration of the Koch curve, A_2, may then be obtained by applying the same four affine transformations to A_1. Higher-order versions of the Koch curve are generated by simply repeating the iterative process until the desired resolution is achieved.

In order to further illustrate this important point, the affine transformation matrices for a Koch fractal curve, a Sierpinski gasket, and a fractal tree, have been provided in Figure 4.6 (a2), (b), and (c), respectively [1, 13].

The first four iterations of the Koch curve are shown in Figure 4.7(a).

We note that these curves would converge to the actual Koch fractal, represented by A_∞ as the number of iterations approaches infinity.

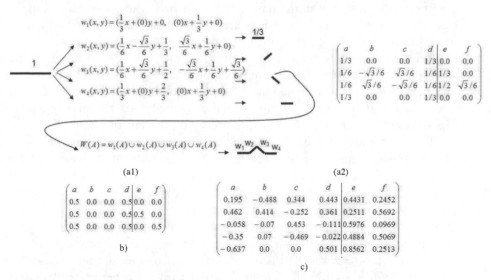

(a1)

(a2)

$$\begin{pmatrix} a & b & c & d & e & f \\ 0.5 & 0.0 & 0.0 & 0.5 & 0.0 & 0.0 \\ 0.5 & 0.0 & 0.0 & 0.5 & 0.5 & 0.0 \\ 0.5 & 0.0 & 0.0 & 0.5 & 0.0 & 0.5 \end{pmatrix}$$

b)

$$\begin{pmatrix} a & b & c & d & e & f \\ 0.195 & -0.488 & 0.344 & 0.443 & 0.4431 & 0.2452 \\ 0.462 & 0.414 & -0.252 & 0.361 & 0.2511 & 0.5692 \\ -0.058 & -0.07 & 0.453 & -0.111 & 0.5976 & 0.0969 \\ -0.35 & 0.07 & -0.469 & -0.022 & 0.4884 & 0.5069 \\ -0.637 & 0.0 & 0.0 & 0.501 & 0.8562 & 0.2513 \end{pmatrix}$$

c)

Figure 4.6 The standard Koch curve as an iterated function system, IFS (a1), and the affine transformation matrices for a Koch fractal curve (a2), a Sierpinski gasket (b), and for a fractal tree (c) [1]. (After [1].)

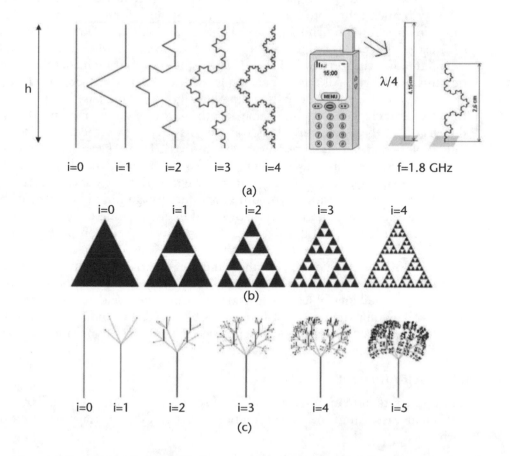

Figure 4.7 (a) The first stages in the construction of the standard Koch curve, (b) the Sierpinski gasket, and (c) the fractal tree via an iterated function system (IFS) approach.

Iterated function systems have proven to be a very powerful design tool for fractal antenna engineers. This is primarily because they provide a general framework or the description, classification, and manipulation of fractals.

Fractal geometries are generated in an iterative fashion, leading to self-similar structures. This iterative generating technique can best be conveyed pictorially, as in Figure 4.7. The starting geometry of the fractal, called the initiator, depends of final fractal shape: each of the straight segments of the starting structure is replaced with the generator, which is shown on the left of Figure 4.7.

The first few stages in the construction of the Sierpinski gasket are shown in Figure 4.7(b.) The procedure for geometrically constructing this fractal begins with an equilateral triangle contained in the plane, as illustrated in stage $i=0$ of Figure 4.7(b). The next step in the construction process (see stage $i=1$) is to remove the central triangle with vertices that are located at the midpoints of the sides of the original triangle, shown in stage $i=0$. This process is then repeated for the three remaining triangles, as illustrated in stage $i=2$ of Figure 4.7(b). The next two stages (i.e., $i=3$ and 4) in the construction of the Sierpinski gasket are also shown in Figure 4.7(b). The Sierpinski-gasket fractal is generated by carrying out this iterative process an infinite number of times. It is easy to see from this definition that the Sierpinski gasket is an example of a self-similar fractal.

The fractal tree, shown in Figure 4.7(c), is similar to a real tree, in that the top of every branch splits into more branches. The planar version of the tree has the top third of every branch split into two sections. The three-dimensional version (this use of the term is only meant to imply that the structure cannot be contained in a plane) has the top third of each branch split into four segments that are each one-third in length. All the branches split with 60° between them. The length of each path remains the same, in that a path walked from the base of the tree to the tip of a branch would be the same length as the initiator. Finding the fractal dimension of these structures is not as easy as it is to find the dimension of the self-similar fractals that were previously observed. This is because the tree fractal is not necessarily self-similar. Mandelbrot suggests that depending on the constructing geometry, the shape may not truly be fractal in the entire structure.

This iterative generating procedure continues for an infinite number of times. The final result is a curve or area with an infinitely intricate underlying structure that is not differentiable at any point. The iterative generation process creates a geometry that has intricate details on an ever-shrinking scale. In a fractal, no matter how closely the structure is studied, there never comes a point where the fundamental building blocks can be observed. The reason for this intricacy is that the fundamental building blocks of fractals are scaled versions of the fractal shape. This can be compared to it not being possible to see the ending reflection when standing between two mirrors. Closer inspection only reveals another mirror with an infinite number of mirrors reflected inside.

4.2.2 Self-Affine Sets

Self-affine sets form an important class of sets, which include self-similar sets as a particular case. An affine transformation $S: \Re^n \to \Re^n$ is a transformation of the form [7]

$$w(x, y) = T(x, y) + t \tag{4.26}$$

where T is a linear transformation on $\Re^n$ (which may be represented by an $n \times n$ matrix) and t is a vector in $\Re^n$. Thus an affine transformation w is a combination of a translation, rotation, dilation, and perhaps reflection (see Figure 4.5). In particular, w maps spheres to ellipsoids, squares to parallelograms, and so forth. Unlike similarities, affine transformations contract with differing ratios in different directions.

If $w_1,...,w_m$ are self-affine contractions on $\Re^n$, the unique compact invariant set F for the w_i is termed a self-affine set. An example is given in Figure 4.5: w_1, w_2, and w_3 are defined as the transformations that map the square onto the three rectangles in the obvious way.

It is natural to look for a formula for the dimension of self-affine sets that generalizes formula for self-similar sets. We would hope that the dimension depends on the affine transformations in a reasonably simple way, easily expressible in terms of the matrices, and vectors that represent the affine transformation. Unfortunately, the situation is much more complicated than this. If the affine transformations are varied in a continuous way, the dimension of the self-affine set need not change continuously. With such discontinuous behavior, which becomes worse for more involved sets of affine transformations, it is likely to be difficult to obtain a general expression for the dimension of self-affine sets.

4.3   Deterministic Fractals as Antennas

Having seen the geometric properties of fractal geometry, it is interesting to explain what benefits are derived when such geometry is applied to the antenna field [14]. Fractals are abstract objects that cannot be physically implemented. Nevertheless, some related geometries can be used to approach an ideal fractal that are useful in constructing antennas. Usually, these geometries are called prefractals or truncated fractals. In other cases, other geometries such as multitriangular or multilevel configurations can be used to build antennas that might approach fractal shapes and extract some of the advantages that can theoretically be obtained from the mathematical abstractions. In general, the term fractal antenna technology is used to describe those antenna engineering techniques that are based on such mathematical concepts that enable one to obtain a new generation of antennas with some features that were often thought impossible in the mid-1980s.

After all the work carried out thus far, one can summarize the benefits of fractal technology in the following way:

1. Self-similarity is useful in designing multifrequency antennas, as, for instance, in the examples based on the Sierpinski gasket, and has been applied in designing of multiband arrays.

2. Fractal dimension is useful to design electrically small antennas, such as the Hilbert, Minkowski, and Koch monopoles or loops, and fractal-shaped microstrip patch antennas.

3. Mass fractals and boundary fractals are useful in obtaining high-directivity elements, undersampled arrays, and low-sidelobes arrays.

4.3.1 Fractals as Wire Antenna Elements

Apparently, the earliest published reference to use the terms fractal radiators and fractal antennas to denote fractal-shaped antenna elements appeared in May 1994 [1, 15]. Prior to this, the terminology had been introduced publicly during an invited IEEE seminar held at Bucknell University in November 1993 [16]. The application of fractal geometry to the design of wire antenna elements was first reported in a series of articles by Cohen [17–20]. These articles introduce the notion of fractalizing the geometry of a standard dipole or loop antenna. This is accomplished by systematically bending the wire in a fractal way, so that the overall arc length remains the same, but the size is correspondingly reduced with the addition of each successive iteration. It has been demonstrated that this approach, if implemented properly, can lead to efficient miniaturized antenna designs. For instance, the radiation characteristics of Minkowski dipoles and loops were originally investigated in [17–20]. Properties of the Koch fractal monopole were later considered in [21, 22]. It was shown that the electrical performance of Koch fractal monopoles is superior to that of conventional straight-wire monopoles, especially when operated in the small-antenna frequency regime. Monopole configurations with fractal top-loads have also been considered, as an alternative technique for achieving size miniaturization.

A fractal can fill the space occupied by the antenna in a more effective manner than the traditional Euclidean antenna [6]. This leads to more effective coupling of energy from feeding transmission lines to free space in less volume. Fractal loop and fractal dipole wire radiators are contrasted with linear loop and dipole antennas, allowing conclusions to be drawn about the effectiveness of the space-filling concepts and correlating the fractal dimension to the miniaturization. Fractal antennas do not need to be limited to only wire antennas. However, wire antennas provide a medium that is easy to analyze and fabricate. By looking at the antenna at various stages of its fractal growth, some conclusions can be drawn about how the antenna is operating by observing trends.

4.3.1.1 Fractal Loop Antennas

The space-filling abilities of fractals fed as loop antennas can exhibit two benefits over Euclidean antennas [6]. The first benefit is that the increased space-filling ability of the fractal loop means that more electrical length can be fitted into a smaller physical area. The increased electrical length leads to a lower resonant frequency, which effectively miniaturizes the antenna. The second benefit is that the increased electrical length can raise the input resistance of a loop antenna when it is used in a frequency range as a small antenna. It can be shown that the resistance increase resulting from the increased wire length for a material with a finite conductivity is insignificant in relationship to the miniaturization of the antenna. Miniaturization of a loop antenna using fractals was shown in [17–19]. Here, it

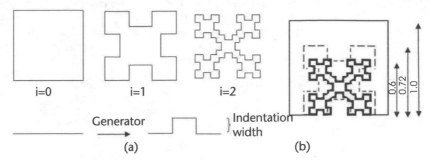

Figure 4.8   (a) The iterative-generation procedure for a Minkowski island fractal, and (b) relative sizes of the loop geometries for square and the first two fractal iterations with an indention width of 0.8, all of the same resonant frequency.

was observed that the resonant frequency of the loop decreased as the generating iterations were increased. The fractal that was used to demonstrate this miniaturization of antennas was a Minkowski square loop. The generation procedure for the fractal is depicted in Figure 4.8.

The initiator is a square, and the generator is depicted in the Figure 4.8. The fractal is formed by displacing the middle third of each straight segment by some fraction of 1/3. This is called the indentation width, w, which can vary from 0 to 1. The resulting structure has five segments for every one of the previous iteration, but not all of the same scale. The fractal dimension can be determined by using the formula [6]

$$N_1(\frac{1}{\delta_1})^{DS} + N_2(\frac{1}{\delta_2})^{DS} + \cdots + N_n(\frac{1}{\delta_n})^{DS} = 1 \qquad (4.27)$$

where N_n is the number of copies of the initiator scaled by δ_n. The subscripts are required to differentiate between the different scales present in the generating iterations. For this example, $N_1=3$ and $\delta_1=3$, representing the horizontal segments, and $N_2=2$ and $\delta_2=3/w$, representing the vertical segments in the generator. For closed fractals that are formed by replicating the structure around the sides of a shape, the fractal dimension needs to be calculated based on the geometry of one side only, since the self-similarity is valid for straight segments and not the entire square. The fractal dimensions of the Minkowski fractal for various values of the indentation width w and number i of generating iterations are tabulated in Table 4.2 [6].

It can be seen how the miniaturization increases as the fractal dimension increases [6]. The perimeter C length of the fractal is given by the following equation:

$$C_i = (1 + \frac{2}{3}w)^i \cdot C_{i-1} \qquad (4.28)$$

where i is the number of generating iterations, and w is a indentation width.

Table 4.2   The Minkowski Fractal Dimension, Perimeter, and Height for the Various Cases of
Indentation Widths and Generating Iterations [6]

Indentation Width Scaling Factor	Generating Iterations	Fractal Dimensions	Perimeter at Resonance [λ]	Height at Resonance [λ]
0	0	1.0	1.11	0.2775
0.2	1	1.0982	1.2149	0.268
0.2	2	1.0982	1.5372	0.264
0.333	1	1.1562	1.2432	0.2543
0.333	2	1.1562	1.798	0.2462
0.5	1	1.228	1.2688	0.2379
0.5	2	1.228	2.1239	0.224

Because the term inside the parentheses is greater than one, the perimeter of
the fractal grows exponentially with increasing iterations (Table 4.2).

The resonant frequencies of the fractal loops were calculated using the
Moment Method [6] at each generating iteration and using various indentation
widths. It can be seen that the resonant frequency of the loop antennas decreases
as the number of generating iterations are increased, or as the indentation width is
increased. The resulting heights of the loop antennas are tabulated in Table 4.2
and plotted in Figure 4.9.

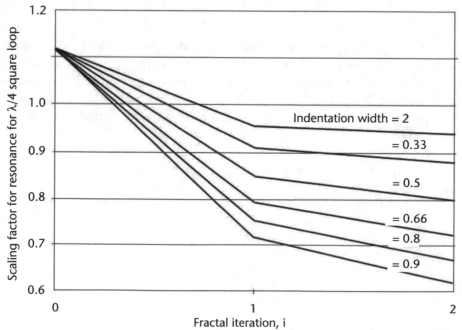

Figure 4.9   Design curves for predicting the resonant frequency of a Minkowski fractal loop for
various numbers of generating iterations and for varying indentation widths. (From:
[6]. ©2002 IEEE. Reprinted with permission.)

Two conclusions can be drawn from the analysis of the Minkowski loop antennas. The first is that the perimeter is not the only factor in determining the first resonance of the antenna. It can be seen that the perimeter increases at a rate faster than the decreasing resonant frequency. This is the expected result, since it is known that the perimeter of the fractal goes to infinity for an ideal fractal, and that the lowering of the resonant frequency saturates after a few generating iterations. The second conclusion that can be drawn is that the resonant frequency of the antenna decreases as the fractal dimension increases. This is clearly the case when the indentation width is increased. Also, the dimension of the prefractal can generally be assumed to be the same as the ideal fractal, if enough iterations are used in the generation and if the increased intricacy is not disenabled. It can be assumed that as the number of generating iterations is increased, the dimension increases from a one-dimensional line to the dimension calculated of the ideal fractal. Thus, the increase in the iterations increases the fractal dimension, which lowers the resonant frequency. Two iterations of the loop were fabricated [6] to be resonant at the same frequency, and measured. They are fed at the corner via coax through a bazooka balun. The scaling factor of the indentation width is 0.5 for the fabricated geometry. This geometry exhibits a 15% reduction in height from the square loop to the prefractal loop at identical resonances. The computed and measured input matches, and far-field patterns in the principal planes are plotted in Figure 4.10 [6].

The shift in resonant frequency between the computed and measured results can be attributed to manufacturing tolerances. However, there is good agreement between the simulated and measured bandwidths. The patterns of the square loop and the 15%-smaller fractal loop are very similar. The increase in the cross-polarization levels of the fractal loop compared with the square loop can be attributed to the jaggedness of the structure, which varies the direction of the current. Furthermore, the radiating untuned balun can lead to increased cross-polarization levels from the measured results, compared to the computed patterns.

An additional facet of the reduction in size of the resonant antenna is that the input resistance of small loops can be increased using fractals. One problem with a small loop is that the input resistance is very low, making it difficult to couple power to the antenna. By using a fractal loop, the antenna can be brought closer to resonance, which raises the input impedance. A Koch island was used as a loop antenna to increase the input resistance. A Koch island (mathematical fractal: the Koch snowflake) is generated (see Figure 4.11) by forming a Koch curve on the three segments of an equilateral triangle in two steps: step one—start with a large equilateral triangle, and step two—make a star by dividing one side of the triangle into three parts and remove the middle section, replacing it with two lines the same length as the section removed, and doing this to all three sides of the triangle. This process should be repeated infinitely. The snowflake has a finite area bounded by a perimeter of infinite length! After the first step, we get four segments (it is then divided into four parts). The whole curve is composed of three of these new segments, so the fractal dimension is DS=log 4/log 3=1.2618. It takes more space than a one-dimensional line segment, but it occupies less space than a filled two-dimensional square.

The process is similar to the formation of the Minkowski fractal loop, except that the generator is comprised of only four segments of equal length, instead of

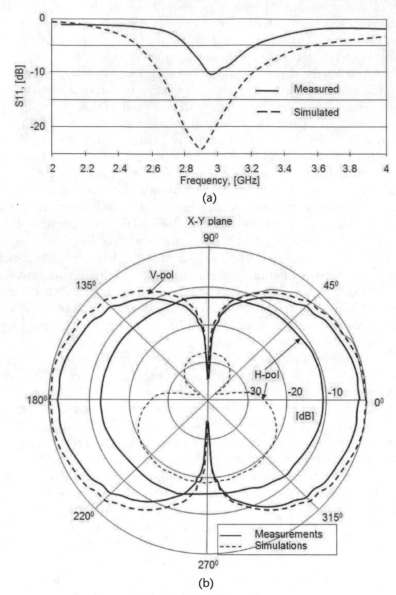

Figure 4.10   The input match (a), and the measured and computed far-field patterns of the first-order Minkowski fractal loop antenna, xy plane (b). (From: [6]. ©2002 IEEE. Reprinted with permission.)

five segments of two different scales. For comparison, a circular loop with equal radius was also constructed, which circumscribes the fractal loop.

Numerical simulations of the circular loop and Koch fractal loop antennas predicted an input resistance of 1.17Ω for the circular loop, and 26.7Ω for the fourth iteration of the Koch loop [6]. The perimeter lengths of the circular and fractal loops were 0.262λ and 0.682λ, respectively. Even though the loops require the same space, it is much easier to couple power to the fractal loop.

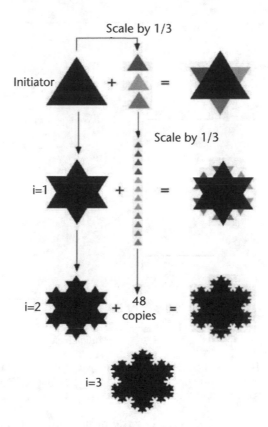

Figure 4.11   The three generating iteration of the Koch loop (snow flake).

4.3.1.2   Fractal Dipole and Monopole Antennas

An interesting study of the space-filling properties of fractal antennas is to investigate fractal dipole antennas. In a way similar to how the loop antennas were studied in the previous section, this section will include a systematic look at the various geometries, with varying fractal dimensions at various stages of growth.

Three types of fractals were compared as dipoles. They are depicted in Figure 4.7(a) and 4.7(c) for the first stages of growth. They included a Koch curve and a fractal tree. The starting structure for each of the fractals was the same dipole antenna of height h. The Koch dipole has been extensively analyzed in [21–23]. Also, a version of a tree fractal has been studied in [24]. As mentioned in the previous section, the Koch curve is generated by replacing the middle third of each segment with two sides of an equilateral triangle. The resulting curve is comprised of four segments of equal length. As calculated above, the fractal dimension of the Koch curve is 1.2619. The fractal tree is similar to a real tree, in that the top of every branch splits into more branches. The planar version of the tree has the top third of every branch split into two sections. All the branches split with $60°$ between them. The length of each path remains the same, in that a path walked from the base of the tree to the tip of a branch would be the same length as the initiator. The dimension DS of fractal tree can be estimated using equation (4.27). For the planar version of the geometry, $N_1=1$, $\delta_1=3/2$, $N_2=2$, and $\delta_2=3$, resulting

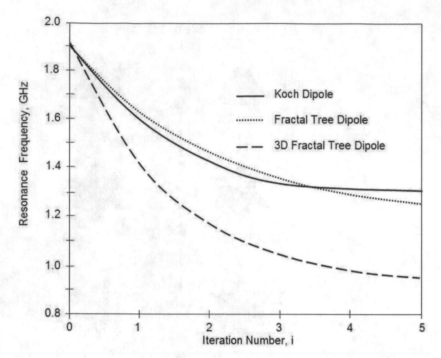

Figure 4.12   The computed resonant frequency for the three types of fractal dipoles as a function
of the number of iterations used in the generating procedure. (From: [6]. ©2002
IEEE. Reprinted with permission.)

in a dimension of $DS=1.395$. For the nonplanar version of the tree, $N_1=1$, $\delta_1=3/2$, $N_2=4$, and $\delta_2=3$, resulting in a dimension of $DS=1.845$.

The resonant frequencies of the antennas, as predicted by the numerical simulations, are plotted in Figure 4.12 [6].

Lowering the resonant frequency has the same effect as miniaturizing the antenna at a fixed resonant frequency. It can be seen how the planar structures have a similar miniaturization response. Switching to the nonplanar three-dimensional version increases this effect. However, it can be seen that the benefits of using fractals to miniaturize are realized within the first several iterations. Along with looking at the resonant frequency of this antenna, it is also quite interesting to look at the quality factor for these antennas.

4.3.1.3 Quality Factor of Fractal Antennas

The quality factor, Q, of an antenna is a measure of the ratio of its radiated energy to its stored energy. For antennas with a $Q \gg 1$, the quality factor is the inverse of the fractional bandwidth. In [25], it states that "it is calculated that the bandwidth of an antenna (which can be closed within a sphere of radius r) can be improved only if the antenna utilizes efficiently, with its geometrical configuration, the available volume within the sphere." Therefore, since fractal geometries efficiently fill spaces, it should be worthwhile to consider the fractional bandwidth or quality factor of some fractal antennas [6].

The theoretical limit for any antenna, regardless of its shape, is given in [10, 26] as

$$Q = \frac{1 + 2(kr)^2}{(kr)^3 [1 + 2(kr)^2]} \tag{4.29}$$

where r is the radius of sphere that completely encloses the antenna, and k is the wavenumber.

Following the procedure for calculating the Q of a single-port antenna given by Puente et al. in [10], the quality factor is a measure of comparing the stored electric or magnetic energy to the radiated energy,

$$Q = \omega \frac{2W_e}{P_r}, \quad W_e > W_m$$
$$Q = \omega \frac{2W_m}{P_r} \quad W_e < W_m \tag{4.30}$$

The stored magnetic and electric energy (i.e., W_m and W_e) can be related to the input impedance of a lossless one-port network by the following relation, given in [26] by

$$W_e = \frac{|I|^2}{8} \left(\frac{dX_{in}}{d\omega} - \frac{X_{in}}{\omega} \right) = \frac{|V|^2}{8} \left(\frac{dB_{in}}{d\omega} - \frac{B_{in}}{\omega} \right)$$
$$W_m = \frac{|I|^2}{8} \left(\frac{dX_{in}}{d\omega} - \frac{X_{in}}{\omega} \right) = \frac{|V|^2}{8} \left(\frac{dB_{in}}{d\omega} - \frac{B_{in}}{\omega} \right) \tag{4.31}$$

Since I and V are the current and voltage at the input terminals of the antenna,

$$P_L = \frac{1}{2} |I|^2 R_{in} = \frac{1}{2} |V|^2 G_{in} \tag{4.32}$$

and, therefore, the quality factor can be computed as

$$Q = \frac{\omega}{2R_{in}} \left(\frac{dX_{in}}{d\omega} + \left| \frac{X_{in}}{\omega} \right| \right) \tag{4.33}$$

Computing the quality factor of fractal wire-dipole antennas can show how fractals fill space in a more efficient manner than linear dipoles, and therefore have a lower Q. Figure 4.13 shows the quality factor of Koch fractal wire dipoles [6].

The Q factors were computed using Moment Method by analyzing the input impedance over a range of frequencies in (4.33). The Q decreased as the number iterating generations was increased for the fractals, as would be expected. Each increasing generated iteration brings the geometry away from the linear, one-dimensional dipole, and closer to the ideal fractal. Also, it can be seen that the three-dimensional fractal tree had the lowest Q of the fractals studied, showing that this antenna is the most effective at filling space. The zero-order fractal, a lin-

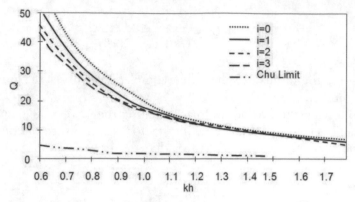

Figure 4.13   A plot of the computed quality factor Q of the Koch dipole as a function of *kh*. (From: [6]. ©2002 IEEE. Reprinted with permission.)

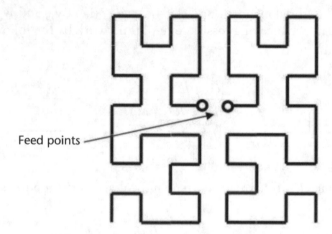

Figure 4.14   Antenna configuration with Hilbert curve fractal patterns.(From: [27]. ©2003 John Wiley & Sons. Reprinted with permission.)

ear dipole, is plotted in Figure 4.13 for comparison, along with the fundamental Chu's limit.

Antenna dipole configuration with Hilbert curve fractal patterns for first three iterations is shown in Figure 4.14 [27].

Hilbert curves are plane-filling curves, apart from being self-similar and simple. The plane filling nature of this geometry can be interpreted as the cause for their relatively lower resonant frequency. It is clear that, as the fractal iteration order increases, the total length of the line segments contributing to the geometry increases in almost geometric progression, even as the area it encompasses remain the same. Thus within a small area, a resonant antenna with very large line length can be accommodated. An approximate formulation for the resonant frequency of a dipole Hilbert curve fractal antenna is derived in [11]. The first few resonant frequencies of it can be obtained from:

$$m \frac{\eta}{\pi \omega} \log \frac{2d}{b} \tan kd + \frac{\mu_0}{\pi} s (\log \frac{8s}{b} - 1) = \frac{\mu_0}{\pi} \frac{n\lambda}{4} (\log \frac{2n\lambda}{b} - 1) \quad (4.34)$$

where η is the intrinsic impedance of free space, ω is angular frequency, λ is the wavelength, k is the wavenumber, n is an odd integer and d, m, s are expressed in terms of fractal iteration order i, $d = l/(2^i - 1)$ is the length of each line segment, l is the outer dimension of Hilbert curve fractal antenna of order i, $m = 4^{i-1}$ are short-circuited parallel wire sections, each of length d and diameter b, and the segments not forming the parallel wire sections amount to a total length of $s = (2^{2i-1} - 1)d$.

This expression does not account for higher order effects and hence may not be accurate at higher resonant modes. The formulas presented here can be appropriately inverted to obtain the design equations for the antenna, for a given resonant frequency.

It would be interesting to identify the fractal properties of this geometry [28]. The plane-filling nature is evident by comparing the first few iterations of the geometry shown in Figure 4.4. It may, however, be mentioned that this geometry is not strictly self-similar since additional connection segments are required when an extra iteration order is added to an existing one. But the contribution of this additional length (shown with dashed lines in Figure 4.4) is small compared to the overall length of the geometry, especially when the order of the iteration is large. Hence, this small additional length can be disregarded, which makes the geometry self-similar. A similar ambiguity also prevails in determining the dimension of the geometry. The topological dimension of the curve is 1 since it consists only of line segments. However, the dimension of a fractal curve can be defined in terms of a multiple-copy algorithm. The similarity dimension DS of the Hilbert curve of order $i \geq 2$ is [28]

$$DS = \frac{\log[(4^i - 1)/(4^{i-1} - 1)]}{\log[(2^i - 1)/(2^{i-1} - 1)]} \quad (4.35)$$

For large order i curves the similarity dimension of Hilbert fractal approaches an integer value $DS = 2$. But if we consider the length and number of line segments first, second, and third iterations, the dimensions are 1.465, 1.694, and 1.834, respectively. These numbers point to the fact that the dimension of the geometry is still a fractional number, albeit approaching 2. As the dimension approaches 2, the curve is almost filling plane. In other words, the total length of the line segments (with topological dimension 1) tends to be extremely large. This could lead to a significant advantage in antennas since the resonant frequency can be reduced considerably for a given area by increasing the fractal iteration order. It may be recalled that the dimension of this curve is larger than that of Koch curves ($DS = 1.262$) studied elsewhere [29], resulting in a larger reduction factor for the antenna size. The studies presented in [28] indicate that, by increasing the fractal iteration order, the resonant frequency of the antenna can be significantly reduced. Thus, this approach strives to overcome one of the fundamental limitations of antenna engineering with regard to small antennas [26]. It may, however, be noticed that, since fractals do not come under the purview of Euclidean geometry, stipulations based on this may be relaxed for fractals [10]. The length L for the Hilbert monopole increases at each iteration i, given by [27]

$$L(i) = \frac{4^{i+1} - 1}{2^{i+1} - 1} h$$ (4.36)

where i is the iteration order, and h is the high of linear monopole as well as the side length of Hilbert fractal. For the fifth iteration, the length is 65 times h—the high of the linear monopole. One might think that the resonant frequency for a fifth iteration Hilbert monopole could be 65 times less than the resonant frequency of a linear monopole with height h, which will be an extraordinary frequency reduction. Obviously, this is not true since coupling between turns provides a shorter path for currents flowing from one tip to the other. However, even with the coupling effect, the resonant frequency reduction or compression efficiency (CE, introduced by [27], is defined as the ratio between the first resonant frequency of the equivalent vertical monopole with a height equal to the total length of a fractal monopole and the first resonant frequency of the fractal antenna) can achieve values up to 10, which is a very significant size reduction. A large CE is of interest for the design of miniature antennas, where space is a constraint factor, such as integrated low-frequency antennas in MMIC circuits. Table 4.3 summarizes the resonant frequencies, the electrical size as well as the CE [27].

The electrical size for the linear monopole does not match the theoretic $\lambda/4$ because the dielectric support shifts the frequency to lower values. The CE increases as the fractal iteration increases, achieving CE=11 for the fifth Hilbert iteration. This means that for the same frequency, using the Hilbert fifth monopole antenna one can achieve a size 11 times less than the size of the classical $\lambda/4$ monopole. It should be noted that the maximum CE achieved for the Koch monopole is CE=1.57, which is a very poor value compared with the Hilbert monopole

Table 4.3   Performance Evolution as a Function of the Fractal Iteration [27]

Antenna		Hilbert		Koch	
	Resonant Frequency [MHz]	Electrical Size [λ]	CE Hilbert	Electrical Size [λ]	CE Koch
Fractal iteration i					
1	236.4	0.055	3.38	0.144	1.21
2	161.1	0.037	4.96	0.129	1.36
3	118.7	0.027	6.73	0.117	1.50
4	86.3	0.020	9.26	0.112	1.56
5	71.7	0.016	11.1	0.111	1.57
Monopole	799.6	0.186	—	—	—

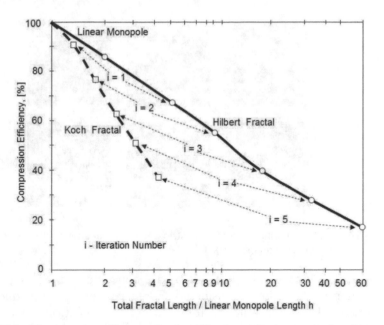

Figure 4.15   Compression efficiency for the Hilbert and Koch monopoles. (From: [14]. ©2005 John Wiley & Sons. Reprinted with permission.)

[21]. The large CE obtained with the Hilbert monopole can be related to the fractal dimension. The fractal dimension for the Hilbert and Koch monopoles are $DS=2$ and 1.26, respectively. This larger fractal dimension indicates that the Hilbert curve fills a surface more efficiently than the Koch curve, which, although it has not been theoretically proved yet, seems to be related to the miniaturization capabilities of fractal-shaped antennas.

CE is an important parameter for comparing different space-filling geometries in order to determine which structure can decrease the resonant frequency with less length.

Figure 4.15 shows the CE as a function of the fractal iteration showing that CE decreases as the fractal iteration increases [27]. In Figure 4.15, the CE for the Koch monopole is depicted [21]. It is clearly shown that the CE decreases more rapidly for the Koch monopole. For example, for the same height h of the vertical monopole, the Hilbert 1 and the Koch 5 monopole have approximately the same length; however, the CE is 70% for the Hilbert monopole and only 40% for the Koch monopole, which shows that the Hilbert monopole achieves a larger frequency reduction within the same wire length.

4.3.2   Fractal Patch Elements

Borja and Romeu [30] proposed a design methodology for a multiband Sierpinski microstrip patch antenna. A technique was introduced to improve the multiband behavior of the radiation patterns by suppressing the effects of high-order modes. Finally, high-directivity modes in a Koch-island fractal patch antenna were studied in [31]. It was shown that a patch antenna with a Koch fractal boundary exhibits localized modes at a certain frequency above the fundamental mode, which can lead to broadside directive patterns [1].

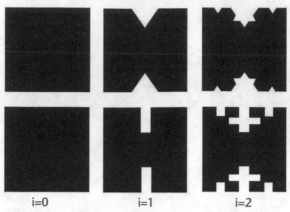

i=0   i=1   i=2

Figure 4.16   Fractal modifications of microstrip patch antenna having the same area for different resonant frequencies ($f_0 > f_1 > f_2$).

Some additional applications of fractal concepts to the design of microstrip-patch antennas have been considered. For instance, [32] introduced a modified Sierpinski-gasket patch antenna for multiband applications. A design technique for bowtie microstrip-patch antennas, based on the Sierpinski-gasket fractal, was presented in [33]. The radiation characteristics of Koch-island fractal microstrip patch antennas were investigated in [34]. Still other configurations for miniaturized fractal patch antennas were reported by Gianvittorio [35], Rahmat-Samii [36], and Krzysztofik [37].

Fractals can be used to miniaturize patch elements as well as wire elements. The same concept of increasing the electrical length of a radiator can be applied to a patch element [6]. The patch antenna can be viewed as a microstrip transmission line. Therefore, if the current can be forced to travel along the convoluted path of a fractal instead of a straight Euclidean path, the area required to occupy the resonant transmission line can be reduced. This technique has been applied to patch antennas in various forms. The torn-square fractal is used along the edge that determines the resonant length in a rectangular patch. The patches are shown in Figure 4.16.

The generating methodology is very similar to that of a Koch curve. The straight radiating edges of the patch are held constant in width while being brought closer to each other. This keeps the gain at the same level for the resonant linearly polarized patch as for the resonant, linearly polarized rectangular patch.

Both of the patches are the same width, 30 mm, while the length of the fractal patch is 38% shorter. The input match is shown in Figure 4.17 [6].

The measured results, as well as the Moment Method simulated results generated with Ansoft's Ensemble, are shown. It can be seen that the fabrication process shifted the resonant frequency higher by 4%, to 5.2 GHz. It is interesting to note from the input match that the price to be paid for miniaturization with this technique is bandwidth. The bandwidth of the resonant rectangular patch is 1.8%, while the bandwidth of the fractal patch is only 0.4%. However, they are both very narrowband, and this may be a worthwhile trade-off for a particular application. The current distribution on the fractal patch as computed by the

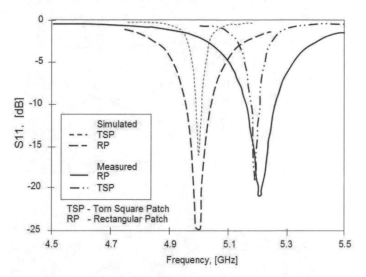

Figure 4.17   The measured and simulated input matches for fractal and rectangular patches. (From: [6]. ©2002 IEEE. Reprinted with permission.)

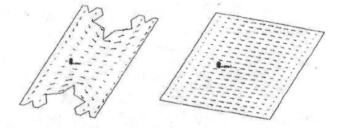

Figure 4.18   The computed current distribution on the surface of a resonant torn-square fractal patch and a rectangular patch. (From: [6]. ©2002 IEEE. Reprinted with permission.)

Moment Method is shown in Figure 4.18, along with the current distribution on a resonant rectangular patch [6].

Figure 4.18 shows how the fractal boundary forces the current to take a circuitous path, therefore lowering the required length for resonance. The length of the resonant rectangular patch is 19.46 mm, while the length of the fractal patch is only 12 mm. However, the length of the boundary, if stretched out straight, is 30.17 mm. It can be seen from Figure 4.18 that the current does not strictly follow the path of the edge, but is rather nudged around it in a manner such as a fast car passing several jutting obstacles. The measured far-field patterns are shown in Figure 4.19 [6].

It can be seen that the patterns of the resonant rectangular patch and the resonant fractal patch are very similar, including the levels of cross polarization. The only difference is a slight narrowing of the H-plane for the fractal patch.

4.3.2.1   The Sierpinski Gasket

One of the fractal structures was discovered in 1916, by Polish mathematician Waclaw Sierpinski. The Sierpinski sieve of triangles possesses a certain multiband behavior owing to its self-similar shape, where a monopole antenna based of the

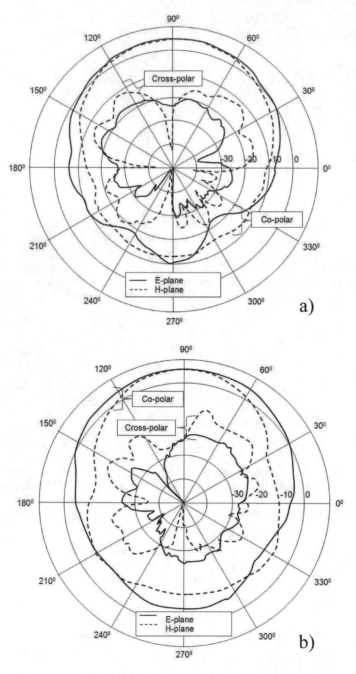

Figure 4.19 The measured far-field gain patterns for the resonant rectangular patch (a), and for the resonant fractal patch (b) antennas. (From: [6]. ©2002 IEEE. Reprinted with permission.)

Sierpinski gasket has been shown to be an excellent candidate for multiband applications. A multiband fractal monopole antenna [1], based on the Sierpinski gasket, was first introduced by Puente et al. [38]. The original Sierpinski monopole antenna is illustrated in Figure 4.20(a).

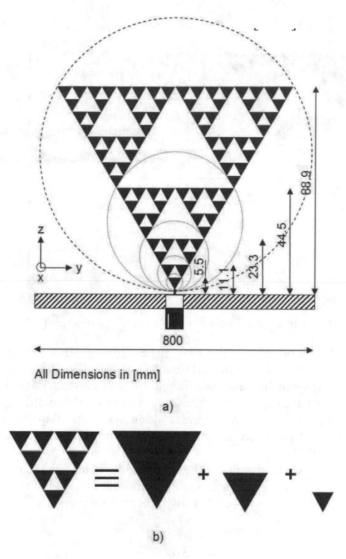

Figure 4.20   Dimensions of a five-iteration Sierpinski fractal monopole (a), and second iteration Sierpinski gasket as equivalent to an array of three triangular antennas (b). (From: [1]. ©2003 IEEE. Partially reprinted with permission.)

From an antenna engineering point of view, a useful interpretation of Figure 4.20(a) is that the dark triangular areas represent a metallic conductor, whereas the white triangular areas represent regions where metal has been removed. With a few exceptions (especially the log-periodic), we typically use a single antenna (size) for each application (frequency band) as depicted in Figure 4.20(b), so in this case (for the second iteration, $i=2$) we have the three-in-one compact antenna.

The geometry of Sierpinski gasket fractal antenna is fully determined by four parameters, namely, the height h of triangle, the flare angle α, the iteration number i, and the scaling factor δ. As described in [38], the Sierpinski monopole presents a log-periodic behavior [25, 37, 39] (see Figure 4.21).

With the bands approximately spaced by a factor $\delta = 2$, the antenna keeps a notable degree of similarity through the bands, with a moderate bandwidth

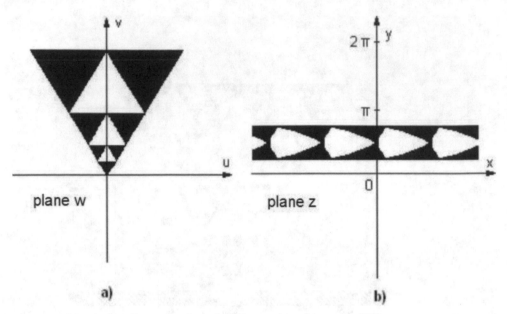

Figure 4.21 (a, b) Affine projection of the Parany fractal structure [37].

(21%) at each one [1]. In this case, the antenna geometry is in the form of a classical Sierpinski gasket, with a flare angle of $\alpha=60°$ and a self-similarity scale factor is determined by ratio of triangle heights of successive iterations, $\delta = \frac{h_i}{h_{i+1}} = 2$. The plots of simulated and measured values of the input reflection coefficient versus frequency for the antenna, along with the associated curves for input resistance and reactance for a prototype Sierpinski gasket monopole are given in Figure 4.22 [1].

To get a better insight on the log-periodic behavior of the antenna, the input impedance is also shown in a semilogarithmic scale in Figure 4.22. It can be seen that the antenna is matched approximately at frequencies

$$f_{r\,i} = 0{,}3 \cdot \cos\left(\frac{\alpha}{2}\right) \cdot \sqrt{\frac{2{,}5}{\varepsilon_r}} \cdot \frac{c}{h} \delta^i \qquad (4.37)$$

where c is the speed of light in vacuum, h is the height of the largest gasket, δ the log period, and i a natural number of iterations, and ε_r is the permittivity of the antenna substrate [38–40]. Such a behavior is clearly different than that of the bowtie monopole, which has the first minimum VSWR at $h/\lambda = 0.17$ and the corresponding higher order modes periodically spaced by a frequency gap of $\Delta f = 0.44c/h$ Hz [40]; that is,

$$f_r = \frac{2c}{3a\sqrt{\varepsilon_r}}\sqrt{m^2 + mn + n^2} \qquad (4.38)$$

where a is the side length of the equally sided triangle, m, n are numbers of EM-field modes generated in the structure, c, and ε_r as in (4.37).

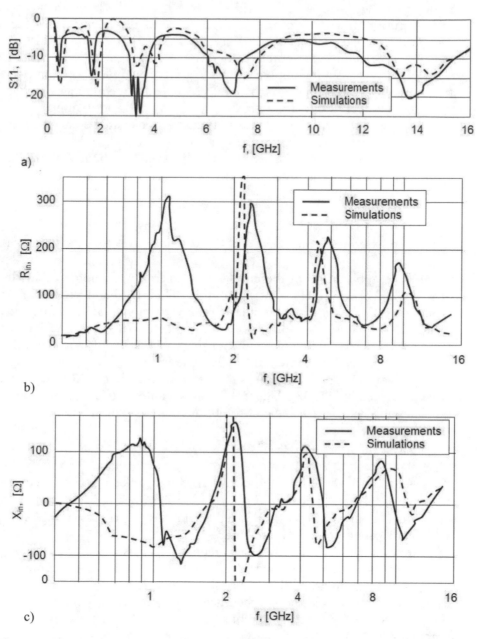

Figure 4.22 The input reflection coefficient (S11) relative to 50Ω (a), the input resistance, R_{in} (b), and the input reactance X_{in} (c) of the five-iteration Sierpinski fractal monopole of Figure 4.21(a). (From: [1]. ©2003 IEEE. Reprinted with permission.)

The second match of the bowtie antenna is always better than the any one of higher mode (S11>15 dB as opposed to 8 dB), which might suggest that it can have a more significant effect on the Sierpinski behavior. Thus, if we assimilate each of the lowest subgaskets to a bowtie, the Sierpinski bands could correspond to the second one of the triangular antennas. The frequency shift toward the origin experimented by the fractal antenna, with respect to the triangular, can be related to the capacitive loading of the upper subgaskets. It is also interesting to notice that the similarity and periodicity are lost in the lower bands where the input return loss and ratio are closer to those of the bowtie. This fact can be related to the antenna truncation since the structure is not an ideal fractal constructed after an infinite number of iterations. Although an ideal fractal shape is self-similar at an infinite number of scales, a feasible implementation of the structure only keeps a certain degree of similarity over a finite number of scales, which limits the number of operating bands. Truncation and end-loading effects of Sierpinski gasket become more apparent when one looks at the current distribution over the antenna and its application to resonant frequency relocation.

A scheme for modifying the spacing between the bands of the Sierpinski monopole was subsequently presented in [38], and later summarized in [1].

4.3.2.2 Variations of Sierpinski Gasket Antennas

It was demonstrated in [1, 38–42] that the positions of the multiple bands may be controlled by proper adjustment of the scale factor used to generate the Sierpinski antenna.

Specific applications of these designs to emerging technologies such as GSM, DECT, WLAN, and UMTS were discussed. The multiband properties of fractal monopoles based on the generalized family of mod-p Sierpinski gaskets were investigated by Castany et al. [43]. The advantage of this approach is that it provides a high degree of flexibility in choosing the number of bands and the associated band spacing for a candidate antenna design.

Dual-band designs, based on a variation of the Sierpinski fractal monopole, were presented in [41, 42]. Figure 4.23 shows an example of a Sierpinski monopole antenna with a flare angle of $\alpha=60°$ and a self-similarity scale factor of

$$\delta = \frac{f_{WLAN2}}{f_{WLAN1}} = \frac{5.8}{2.45} \approx 2.36$$

designed for WLAN terminal applications. Input reflection coefficient, radiation patterns of such antenna are plotted in Figure 4.24.

A novel configuration of a shorted fractal Sierpinski gasket antenna was presented and discussed in [44] (see Figure 4.25).

The multitriangular monopole antenna, which is a variation of the Parany antenna, was originally considered in [45]. These multitriangular antennas have been shown to exhibit multiband properties with respect to input impedance and radiation patterns, even though their geometry is not strictly fractal. In particular, the properties of the Parany antenna are very similar to those of the Sierpinski antenna.

Further investigations concerning enhancing the performance of Sierpinski-gasket monopoles through perturbations in their geometry were reported in [46]. It was found that a variation in the flare angle of the antenna translated into a shift of the operating bands, as well as into a change in the input impedance and radiation patterns.

In order to overcome the problem of miniature microstrip antennas (i.e., small bandwidth and radiation efficiency), parasitic techniques have been combined with fractal techniques to obtain miniature and wideband antennas with improved efficiency [47]. A modified Sierpinski-based microstrip antenna consisting of an active patch and a parasitic patch is presented in [14]. Using such a

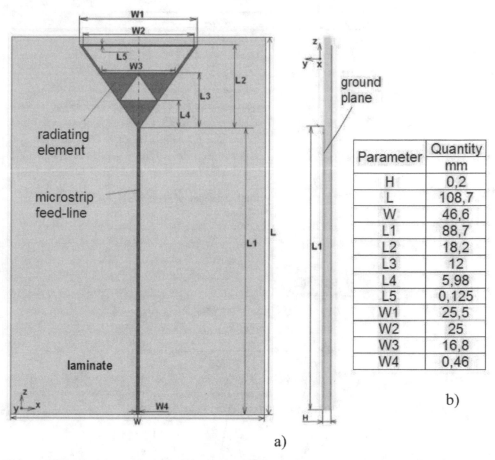

Parameter	Quantity mm
H	0,2
L	108,7
W	46,6
L1	88,7
L2	18,2
L3	12
L4	5,98
L5	0,125
W1	25,5
W2	25
W3	16,8
W4	0,46

a)

Figure 4.23   (a, b) Modified Sierpiski-gasket monopole for WLAN applications [41, 42].

geometry, the resonant frequency of the antenna is 1.26 GHz while it is 2 GHz for the filled version. By adding the parasitic patch, the bandwidth with respect to the single active element is increased by a factor of 15, resulting in a bandwidth of BW=2.7% at SWR=2:1; the radiation efficiency for this antenna is 84%.

4.3.2.3   The Hilbert Curve Meandered Patch Antennas

As it was mentioned many times, an important characteristic of many fractal geometries is their plane-filling nature. Antenna size is a critical parameter because antenna behavior depends on antenna dimensions in terms of wavelength (λ). In many applications, space is a constraint factor, therefore an antenna cannot be comparable to the wavelength but smaller (i.e., a small antenna). An antenna is said to be small when its larger dimension is less than twice the radius of the radian sphere; its radius is $\lambda/2\pi$. Wheeler and Chu were the first who investigated the fundamental limitations of such antennas. The Hilbert and Koch fractal curves have also been useful in designing small microstrip patch antennas. The goal of this section is to present the miniature features of the Hilbert shaped patches and compare it with the Koch shaped patches. Some recent advances are also presented.

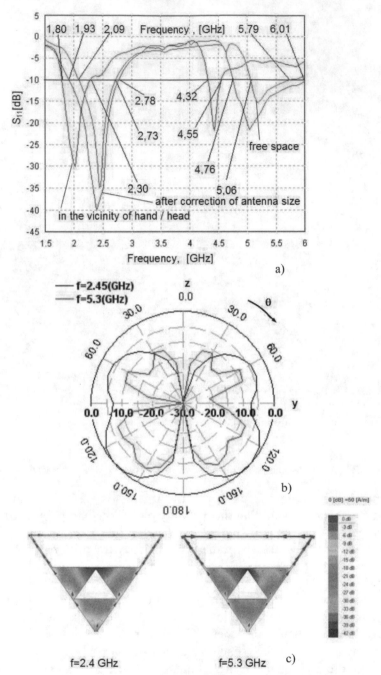

Figure 4.24   (a) Input return loss S11, (b) radiation patterns, and (c) the surface current distribution at the two resonant frequencies of modified Sierpinski-gasket monopole of Figure 4.23.

In [48], the miniature microstrip patch antenna is loaded with a space-filling transmission line based on the fractal Hilbert curve that reduces the resonant frequency of the fundamental mode of the unloaded patch.

Azad et al. [49] have proposed the new PIFA for handset applications, with patch made in the form of Hilbert fractal meander. The antenna geometry con-

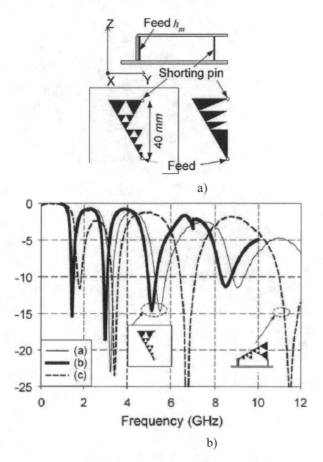

a)

b)

Figure 4.25   (a) Planar shorted configuration of fractal Sierpinski gasket and the Parany monopoles, and (b) the S_{11} characteristics. (From: [44]. ©2004 IEEE. Reprinted with permission.)

forming to the Hilbert profile effectively increases the length of the current flow path, making it possible to develop a miniature antenna. To achieve dual-band performance at least two branches of a radiating element are needed, as shown in Figure 4.26.

For all antenna geometries, a ground plane made of a thin metallic conductor with a length of 110 mm and width of 40 mm was considered. The antenna height was fixed at 10 mm. The study and design of all antennas were conducted using Ansoft HFSS. First, the antenna consisting of two Hilbert elements as shown in Figure 4.26 was considered. The smaller element near the feed point is responsible for resonance in the high frequency band while the larger element is responsible for the resonance in the low frequency band. The models of each element were constructed following a higher-order Hilbert curve. The individual segments and the trace width (0.71 mm) were selected such that the two elements can be accommodated within the width of the ground plane (40 mm). No attempt was made to reduce the trace width and further miniaturize the antenna. All parameters for this antenna are listed in a table inside Figure 4.26. The antenna was studied in the absence of any dielectric material, and it resonated at around 920 and 1,920 MHz. The bandwidths of antenna are 9.4% and 4.4% within 2.5:1 VSWR. The bandwidth in the high frequency band for this configuration is relatively narrow

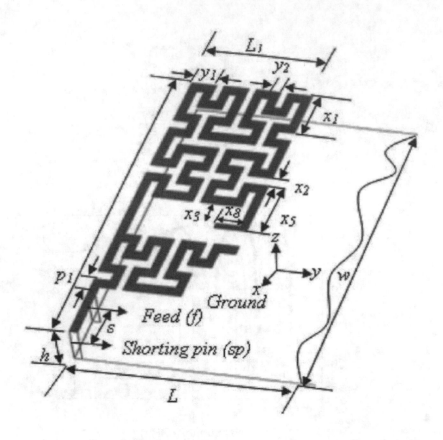

Parameter	x_1	x_3	x_4	y_2	s	L	L_1	W	h
mm	20	2.93	4	0.71	6	110	18.65	40	10

Figure 4.26 Dual-band miniaturized Hilbert PIFA. (From: [49]. ©2005 IEEE. Reprinted with permission.)

for the 1,900-MHz band PCS application. This prompted us to investigate a hybrid geometry consisting of a Hilbert element for the low frequency band and a plate (conventional PIFA) element for the high frequency band.

Computed radiation patterns show the low resonant frequency pattern of uniform coverage in the azimuth plane, which is desirable in mobile wireless application with the component being dominant. In the elevation plane the pattern resembles that of a dipole antenna with some cross-polarization (suppressed below 20 dB). Similarly, the azimuth pattern in the high frequency band is also fairly uniform with both components being close to each other. In the elevation plane the pattern has a butterfly shape. The peak gain at 910 MHz is 2.4 dBi and at 1912 MHz it is 4.6 dBi. A longer antenna and ground plane at the high frequency results in a more directive pattern and, hence, the high gain. It is well known that current on the tip of monopole is zero, which results in zero fields along. But the proposed PIFA has nonzero fields along providing better coverage in the upper hemisphere as compared to vertical monopole antenna. This geometry occupies only about 50% of the volume (only 4.3 cm^3) needed by a conventional metal plate-type PIFA. Computed and measured bandwidth of the antenna indicates good performance for dual-band mobile phone application at around 900 and 1,900 MHz.

4.3.2.4   The Koch Island

The Koch island (or snowflake) is composed of three Koch curves rotated by suitable angles and fitted together or of triangles as depicted in Figure 4.11.

We already know that the length of the Koch curve is immeasurable, so the length of the coastline of the Koch island is seemingly infinite, but what about the area bounded by the perimeter of the island? It certainly looks finite. We can obtain a value for the bounded area by examining the construction process [3]. Let us first assume for simplicity that the initiator is composed of three unit lines. The area bounded by the perimeter is then half of the base multiplied by the height of the equilateral triangle (i.e., $\frac{1}{2} \times 1 \times \frac{\sqrt{3}}{2}$). At step i=1 three smaller triangles are added, each with a base length equal to one third. At step i=2 another twelve smaller triangles are added, each with base length equal to one ninth. At step i=3 forty-eight smaller triangles are added, each with base length of one twenty-seventh. The area then increases, in general, for an arbitrary step i as follows:

$$area = \frac{3\sqrt{3}}{16}\left(\frac{1}{3}\right) + \frac{3\sqrt{3}}{16}\left(1 + \frac{4^1}{9^1} + \frac{4^2}{9^2} + \frac{4^3}{9^3} + ... + \frac{4^i}{9^i}\right) \qquad (4.39)$$

In the limit, as i tends to infinity, the geometric series in the brackets on the right-hand side of the above expression tends to $\left(\frac{9}{5}\right)$, this leaves us with an area of $\left(\frac{2}{5}\sqrt{3}\right)$. The Koch island therefore has a finite area of $\frac{2}{5}\sqrt{3}$, or about 0.693 units (of area). Thus, the Koch island has a regular area, in the sense that it is bounded and measurable, but an irregular, immeasurable perimeter. To generate the Koch island, we used three Koch curves with unit initiator. However, if the initiator was a in length, then the area would be simply $\frac{2}{5}\sqrt{3}a^3$. You can easily verify this for yourself.

Another interesting feature of fractal-based microstrip antennas is the existence of localized modes called fractinos and fractions depending on whether the structure is based on mass fractals or boundary fractals (Figure 4.27) [14].

Figure 4.27 shows a Koch island patch where the current distribution for the fundamental mode and fractino mode are shown. It is worth noting that for the fractino mode there are zones of high current density (localized mode). Moreover, such zones are coherent; that is, they radiate in phase, and thus the radiation pattern is broadside. This feature is rarely obtained with classical Euclidean geometries such as squares, circles, and triangles. It is interesting to note that the directivity of the fractino mode increases with the mode order.

A novel microstrip patch antenna with a Koch prefractal edge and a U-shaped slot is depicted in Figure 4.28.

This combination was proposed by Guterman et al. [50] for multistandard use in GSM1800, UMTS, and HiperLAN2. Making use of an inverted-F antenna (PIFA) structure an interesting size reduction is achieved. As shown in Figure 4.28, the two identical fractal PIFA elements have been disposed to ensure physical symmetry of the antenna structure, which has been useful for MIMO applica-

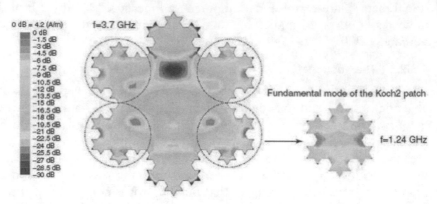

Figure 4.27   Current distribution for the fundamental and localized modes of a Koch microstrip patch. (This figure is available in full color at http://www.mrw.interscience.wiley.com/erfme.) (Reprinted from [14] with permission from John Wiley & Sons)

tions. The antenna has been implemented in microstrip planar technology. A finite ground plane, of dimensions (100×45) mm^2, has been chosen to represent a common handset size. The patch element has been printed on a Duroid 5880 substrate $(\varepsilon_r=2.2, h=1.57$ mm$)$. To meet the bandwidth requirements a 10-mm-thick air gap has been introduced between the substrate and the ground plane. The patch element has been designed based on a simple rectangular shape where the longitudinal resonant edges have been substituted by the fourth iteration of the Koch fractal curve. To obtain an additional miniaturization effect, the Koch-edge patch has been implemented in a PIFA configuration. By doing so, the patch element length has been reduced by 62% in comparison with the simple rectangular patch. The desired upper band (HiperLAN2) antenna behavior has been achieved using a U-shaped slot. The U-shape slot has been cut inside the patch element around the feeding point. For the lower frequency bands (GSM1800 and UMTS), the dimensions of the slot are much smaller than the wavelength, so the cut does not influence the antenna behavior. In this case, the active region covers the entire patch shape, what is well established by surface current distribution, as illustrated in Figure 4.29 [50].

For the HiperLAN2 band the effectively excited area is limited to the interior of the U-shaped slot. In the upper frequency band, the antenna works as a simple rectangular patch. It is not necessary to miniaturize the inner rectangular patch because it fits inside the fractal element. The antenna was simulated in Ansoft Ensemble software tool and manufactured. The antenna is matched, for S11<–6 dB, in the frequency ranges from 1.74 to 2.20 GHz and from 5.07 to 5.4 GHz. Therefore, the required GSM1800, UMTS, and HiperLAN2 bands have almost been covered. The lower antenna band has suffered a shift of 30 MHz, however, the patch can be easily resized to compensate for this small shift. The measured mutual coupling between elements remains below -9 dB in GSM1800, –8 dB in UMTS, and –20 dB in HiperLAN2, which is acceptable for MIMO applications.

4.3.3   Circularly Polarized Microstrip Fractal Antennas

Recently, "open-structure" space-filling curves have been proposed for the development of coplanar waveguide (CPW)-fed fractal slot antennas [51], where

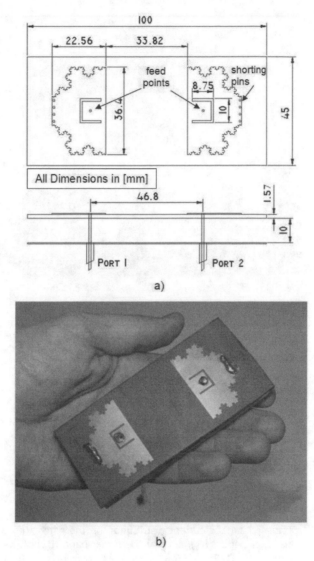

Figure 4.28   (a) Geometry of the microstrip Koch fractal double-PIFA with a U-slot, and (b) in a photo for MIMO application. (From: [50]. ©2004 IEEE. Reprinted with permission.)

miniaturization is the main objective. For this purpose, the conventional slot is replaced by a space-filling curve having the same electrical properties, but confined in a reduced area. The second-iteration Sierpinski slot antenna, operating around 2.4 GHz, presents a compact design confined in (1.8×1.8) cm^2 (0.144×0.144) λ_0^2 with a bandwidth of 5% and a gain of 2.25 dB. On the other hand, the first-iteration Minkowski fractal slot antenna has a bandwidth of 35%, a gain of 5.4 dB, and is confined in (5×5) cm^2.

A circularly polarized fractal slot antenna is proposed in [52]. Slot antennas are useful candidates for integrated planar antennas due to their low cost, ease of manufacture, and flexibility of design in addition to their wideband performance, which is basically required in many recent applications. Currently, recent research studies are going towards miniaturization of slot antennas. One of the promising

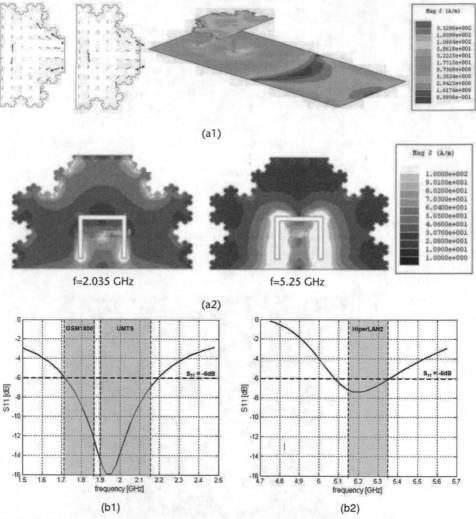

Figure 4.29 Simulated surface current distribution on the patch surface (a1), (a2), and S11 for the lower bands (b1), and for the upper band (b2) of the antenna of Figure 4.28. (From: [50]. ©2004 IEEE. Reprinted with permission.)

techniques for miniaturization is the use of fractal structures in antenna designs. The design is based on the Sierpinski shape. The conductors and nonconductors configurations have been exchanged to produce a Sierpinski slot antenna configuration, reproduced in Figure 4.30.

The first design step is based on the use of a second-iteration Sierpinski gasket. In the second design step, four copies of the Sierpinski slot antenna, with a flare angle of $90°$ are placed at right angles to each other, to form a symmetrical slot antenna in the form of a square. The developed structure represents a new family of fractal slot antennas, where its iteration changes by changing the iteration of the used single-element Sierpinski, as shown in Figure 4.30, where the black sections represent conductors. In addition, by introducing an asymmetry in the proposed slot antenna, which is implemented by extending the ground into the square antenna, circular polarization is achieved. The proposed slot antenna has a 3-dB axial ratio bandwidth of 28% (8.6–11.4 GHz) with VSWR < 2. The antenna

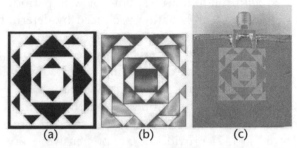

Figure 4.30   (a) CP fractal slot antenna configuration, (b) magnetic current distribution at 11.4 GHz, and (c) fabricated antenna. (From: [52]. ©2005 IEEE. Reprinted with permission.)

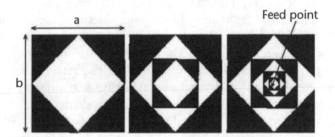

Figure 4.31   Crown-square fractal antenna for circular polarization. (From: [53]. ©2004 IEEE. Reprinted with permission.)

has a maximum gain of 5 dB, radiation efficiency of 97%, and a compact size of (2x2) cm^2. The analysis and design of the proposed antenna have been performed using the full-wave electromagnetic simulator HFSS. The proposed antenna is implemented on RF-35 Taconic substrate (thickness=0.76 mm, ε_r=3.5) with an overall size of (2x2) cm^2, and is fed with 50Ω proximity-coupled microstrip line on the backside of the substrate.

Dehkhoda et al. [53] have proposed a "crown-square" fractal type CP antenna. In Figure 4.31 the first steps of the iteration procedure are shown.

The circular polarization can be achieved by proper adjustment of "square" sides ratio b/a. Authors suggested –b/a=1.046 (a=53.2 mm), and feed point position placed noncentrically, at the 23.36-mm aside of left edge and 24.43 mm above the lower edge of the radiating element.

4.4   Fractal Arrays and Fractal Volume Antennas

The term fractal antenna array was originally coined by Kim and Jaggard in 1986 [54] to denote a geometrical arrangement of antenna elements that is fractal. They used properties of random fractals to develop a design methodology for quasi-random arrays. In other words, random fractals were used to generate array configurations that were somewhere between completely ordered (i.e., periodic) and completely disordered (i.e., random). The main advantage of this technique is that it yields sparse arrays that possess relatively low sidelobes. While this is a feature typically associated with periodic arrays, it is not so for random arrays. Another advantage of the technique is that it is also robust, which in turn is a feature typically associated with random arrays, but not with periodic arrays. The time-harmonic and time-dependent radiation produced by deterministic fractal arrays in the form of Paskal-Sierpinski gaskets was first studied by Lakhtakia et al. [55]. In particular, the radiation characteristics were examined for Paskal-Sierpinski arrays, comprised of Hertzian dipole sources located at each of the gasket nodes. A family of nonuniform arrays, known as Weierstrass arrays, was first introduced in [56]. These arrays have the property that their element spacings and current distributions are self-scalable and can be generated in a recursive fashion.

Werner at al. in [57] provided a comprehensive overview of recent developments in the field of fractal antenna engineering, with particular emphasis placed on the theory and design of fractal arrays. They introduce some important properties of fractal arrays, including the frequency-independent multiband characteristics, schemes for realizing low-sidelobe designs, systematic approaches to thinning, and the ability to develop rapid beam-forming algorithms by exploiting the recursive nature of fractals. These arrays have fractional dimensions that are found from the generating subarray used to recursively create the fractal array. The more general case of fractal planar arrays was analyzed, which was a symmetric planar array of isotropic sources, with elements uniformly spaced a distance d_x, and d_y apart in the x and y directions, respectively. A Sierpinski carpet is a two-dimensional version of the Cantor set, and can similarly be applied to thinning planar arrays. The geometry for this Sierpinski carpet fractal array at various stages of growth is illustrated in Figure 4.32 [57], along with a plot of the corresponding array factor. A comparison of the array factors for the first four stages of construction shown in Figure 4.32 reveals the self-similar nature of the radiation patterns. In this figure, Scale 1 is the generator subarray, Column 2 is the geometrical configuration of the Sierpinski carpet array (white blocks represent elements that are turned on, and black blocks represent elements that are turned off), Column 3 is the corresponding array factor, where the angle ϕ is measured around the circumference of the plot, and the angle θ is measured radially from the origin at the lower left [57].

The multiband nature of the planar Sierpinski carpet arrays may be easily demonstrated by generalizing the argument presented for linear Cantor arrays.

The concept of a fractal volume antenna was introduced in [58], and was demonstrated as a means of increasing the degrees of design freedom for planar fractal antennas, at the expense of some small increase in antenna thickness. Some examples of fractal volume antennas were presented, including a triangular Sierpinski carpet monopole, and a square Sierpinski carpet microstrip antenna. Other examples of fractal volume antennas include the stacked Sierpinski monopole, and the stacked Sierpinski microstrip patch, considered in [59]. The latter approach made use of small parasitically coupled fractal patch elements, in order to increase the bandwidth compared to a single active fractal patch antenna.

An interesting future is in application of 3-D fractal structures: the 3-D Sierpinski triangle, the Mengera Sponge or 3-D Sierpinski carpet, as well as the 3-D Hilbert meander, as shown in Figure 4.33.

These are still waiting for antenna engineers who may apply them, for example, as a base station adaptive antenna array of 4G mobile communication systems.

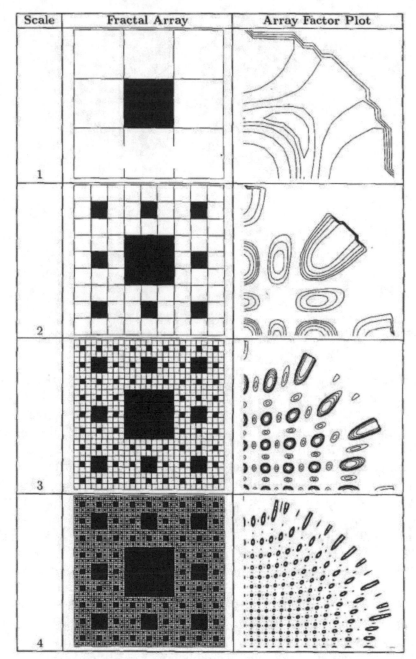

Figure 4.32 The geometry for the Sierpinski carpet fractal planar array (white and black blocks represent elements that are turned on or off, respectively) at various stages of growth, and the corresponding array factor plots. (From: [57]. ©1999 IEEE. Reprinted with permission.)

4.4.1 Multiband Fractal Arrays

A novel application of fractal antenna elements in linear phased arrays was discussed in [6]. The self-similar complexity of a fractal antenna can reduce the overall width of the antenna element at resonance, while maintaining the same

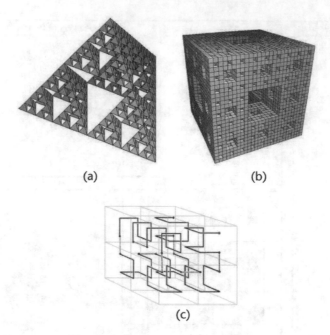

(a)   (b)

(c)

Figure 4.33   (a)The 3-D fractal structures of Sierpinski triangle, (b) the (Karl) Mengera Sponge or
3-D Sierpinski carpet, and (c) Hilbert meander.

performance as a Euclidean element. Because the elements are both resonant and
have similar radiating patterns, the fractal elements can improve designs of linear
arrays that use Euclidean elements. There are two methods using fractal elements
to improve linear arrays that are investigated in [6]. The first method aims to
reduce the amount of mutual coupling between the elements. The second method
packs more elements into a linear array. This concept can be verified with both
wire-loop elements and patch elements. These elements were chosen due to their
similar radiating characteristics. They both radiate with similar patterns, and they
both can be matched at the same frequency. The fractal loop is 30% smaller in
width than a square loop resonant at the same frequency, and the fractal patch is
38% smaller along the resonant length, when compared to a rectangular patch.
Mutual coupling between elements of a linear array can lead to a degradation in
the radiation pattern. Mutual coupling changes the excitation of each element.
Therefore, unless the design procedure of the array includes the effects of mutual
coupling, the radiation pattern will be affected. This is usually manifested by the
raising of sidelobe levels and the filling-in of nulls. The layouts for the arrays used
to compare the effects of mutual coupling between fractal elements and Euclidean
elements are depicted in Figure 4.34 [6, 35, 36].

   Both arrays contain five loop elements, and have a center-to-center spacing of
0.3λ. The elements in both arrays have a progressive phase of 1.632 radians, to
scan the main beam to $\theta=135°$ from the axis of the array. The far-field pattern of
the array—as computed with the Moment Method, which includes all interac-
tions and mutual coupling—is shown in Figure 4.35 [6].

   The far-field patterns of the two arrays are plotted along with the ideal array
factor, which ignores mutual coupling. It can be seen that while both arrays scan

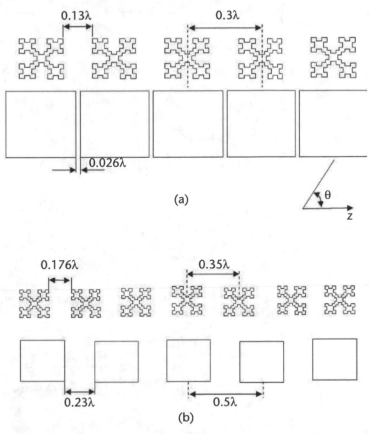

Figure 4.34   (a) The layout of an array with square elements and Minkowski fractals, and (b) the layout of an array with an increased number of fractal elements that fit into the same space as the Euclidean-element array. (From: [6, 35]. ©2002 IEEE. Reprinted with permission.)

the main beam to the desired direction of $\theta=135°$, the array with fractal elements has much less back radiation in the $\theta=45°$ direction. When comparing the far-field pattern of the array with fractal elements to the ideal array factor, it can be seen how mutual coupling has affected the performance and filled in the nulls. However, the degradation of performance is not as severe as it is with the array with square elements. In the $\theta=45°$ direction, the radiation from the array with fractal elements is 20 dB down below that of the array with square elements, in the same direction. This leads to more power directed in the desired direction. Fractals can also be used to pack more resonant elements into a linear array (Figure 4.34(b)). If the overall width of an array is fixed, packing more elements into the same space decreases the center-to-center spacing of the elements. This allows the array to be scanned to lower angles, without grating lobes popping up on the other side. Furthermore, because the fractal elements are smaller at the same resonating frequency, the edge-to-edge spacing between the elements can be maintained. Therefore, this procedure does not increase the amount of mutual coupling between the elements.

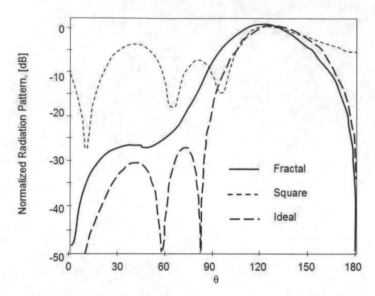

Figure 4.35   The computed far-field pattern of fractal and square loop arrays of Figure 4.34. (From: [6]. ©2002 IEEE. Reprinted with permission.)

Figure 4.36   A flip-up laptop antenna array that can use the tightly packed elements within the fixed width of the screen for wide, efficient scanning in the horizontal plane.

An example of an application that could benefit from this technique is a flip-up linear array that would attach to the screen of a laptop computer, as depicted in Figure 4.36.

This type of scanning antenna could be used to radiate only the surrounding walls of an office, with minimal radiation sent to the floor or ceiling. The techniques presented above can be used to design a highly efficient system.

4.5 Conclusions

When we refer to a fractal, therefore, we will typically have the following in mind: fractal has a fine structure, (i.e., detail on arbitrarily small scales); it is too irregular to be described in traditional geometrical language, both locally and globally; often fractal has some form of self-similarity, perhaps approximate or statistical; usually, the "fractal dimension" is greater than its topological dimension; in most cases of interest fractal is defined a very simple way, perhaps recursively.

Fractal antenna engineering represents a relatively new field of research that combines attributes of fractal geometry with antenna theory. Research in this area has recently yielded a rich class of new designs for antenna elements as well as arrays. Fractals are space-filling geometries that can be used as antennas to effectively fit long electrical lengths into small areas. This concept has been applied to wire and patch antennas. Through characterizing the fractal geometries and the performance of the antennas, it can be surmised that increasing the fractal dimension of the antenna leads to a higher degree of miniaturization. Also, it has been shown that a high degree of complexity in the structure of the antenna is not required for miniaturization. Truncating the fine structure of the fractal that is not discernable at the wavelengths of interest does not affect the performance of the antenna. Therefore, miniaturized antennas can be fabricated using only a few generating iterations of the generating procedure. The applications of these miniaturized elements were demonstrated for phased arrays with enhanced wide-scanning-angle radiation characteristics.

References

[1]   Werner, D. H., and Ganguly, S., "An Overview of Fractal Antenna Engineering Research," *IEEE Antennas and Propagation,* Vol. 45, No. 1, February 2003, pp. 38–57.

[2]   Bamsley, M. F., *Fractals Everywhere,* 2nd ed., New York: Academic Press Professional, 1993.

[3]   Addison, P. S., *Fractals and Chaos: An Illustrated Course,* Bristol, U.K.: Institute of Physics Publishing, 1997.

[4]   Mandlebrot, B. B., *The Fractal Geometry of Nature,* New York: W.H. Freeman, 1983.

[5]   http://www.fractal-landscapes.co.uk/.

[6]   Rahmat-Samii, Y., and Gianvittorio, J. P., "Fractal Antennas: A Novel Antenna Miniaturization Technique and Applications," *IEEE Antennas and Propagation,* Vol. 44, No. 1, February 2002, pp. 20–36.

[7]   Falconer, K. J., *Fractal Geometry: Mathematical Foundations and Applications,* Chinchester, New York: John Wiley & Sons, 1990.

[8]   Best, S. R., and J. D., Morrow, "The Effectiveness of Space-Filling Fractal Geometry in Lowering Resonant Frequency," *IEEE Antennas and Wireless Propagation Letters,* Vol. 1, 2002, pp. 112–115.

[9]   Kritikos, H. N., and Jaggard, D. L., (eds.), *Recent Advances in Electromagnetic Theory,* New York: Springer-Verlag, 1990.

[10] Werner, D. H., and Mittra, R. (eds.), *Frontiers in Electromagnetics,* New York: IEEE Press, 2000.

[11]   Vinoy, K. J., et al., "Resonant Frequency of Hilbert Curve Fractal Antennas," *Proc. IEEE Antennas and Propagation Society Int. Symposium*, Vol. 3, July 2001, pp. 648–651.

[12]   Anguera, J., Puente, C. and Soler, J., "Miniature Monopole Antenna Based on the Fractal Hilbert Curve," *Proc. IEEE Antennas and Propagation Society Int. Symposium*, Vol. 4, June 2002, pp. 546–549.

[13]   Peitgen, H. O., Jurgens, H., and Saupe, D., *Chaos and Fractals: New Frontiers of Science*, New York: Springer-Verlag, 1992.

[14]   Anguera, J., et al., "Fractal-Shaped Antennas: A Review," in *Wiley Encyclopedia of RF and Microwave Engineering*, K. Chang, (ed.), Vol. 2, 2005, pp. 1620–1635.

[15]   Wener, D. H., "Fractal Radiators," *Proc. 4th Annual 1994 IEEE Mohawk Valley Section Dual-Use Technologies & Applications Conference*, Vol. I, SUNY Institute of Technology at Utica/Rome, New York, May 23–26, 1994, pp. 478–482.

[16]   Werner, D. H., "Fractal Electrodynamics," Invited Seminar for the Central Pennsylvania Section of the IEEE, Buchell University, Lewisburg, Pennsylvania, November 18, 1993.

[17]   Cohen, N., "Fractal Antennas: Part I," *Communications Quarterly*, Summer 1995, pp. 7–22.

[18]   Cohen, N., "Fractal and Shaped Dipoles," *Communications Quarterly*, Spring 1996, pp. 25–36.

[19]   Cohen, N., "Fractal Antennas: Part 2," *Communications Quarterly*, Summer 1996, pp. 53-66.

[20]   Cohen, N., and R. G. Hohlfeld, "Fractal Loops and the Small Loop Approximation," *Communications Quarterly*, Winter 1996, pp. 77–81.

[21]   Puente, C., Romeu, et al., "Small But Long Koch Fractal Monopole," *Electronics Letters*, Vol. 34, No. 1, January 1998, pp. 9–10.

[22]   Baliarda, C. P., Romeu, J., and Cardama, A., "The Koch Monopole: A Small Fractal Antenna," *IEEE Transactions on Antennas and Propagation*, Vol. 48, No. 11, November 2000, pp. 1773–1781.

[23]   Best, S. R., "On the Performance of the Koch Fractal and Other Bent Wire Monopoles as Electrically Small Antennas," *Proc. IEEE Antennas and Propagation Society Int. Symposium*, Vol. 4, June 2002, pp. 534–537.

[24]   Werner, D. H., Bretones, A. R., and Long, B. R., "Radiation Characteristics of Thin-Wire Ternary Fractal Trees," *Electronics Letters*, Vol. 35, No. 8, 1999, pp. 609–610.

[25]   Balanis, C. A., *Antenna Theory: Analysis and Design, 2nd ed.*, New York: John Wiley & Sons, Inc., 1997.

[26]   Harrington, R. F., *Time-Harmonic Electromagnetic Fields*, New York: McGraw-Hill, 1961.

[27]   Anguera, J., et al., "The Fractal Hilbert Monopole: A Two-Dimensional Wire," *Microwave and Optical Technology Letters*, Vol. 36, No. 2, January 2003, pp. 102–104.

[28]   Vinoy, K. J., et al., "Hilbert Curve Fractal Antenna: A Small Resonant Antenna for VHF/UHF Applications," *Microwave and Optical Technology Letters*, Vol. 29, No. 4, May 2001, pp. 215–219.

[29]   Cohen, N., "Fractal Antenna Applications in Wireless Telecommunications," *Prof. Program. Proc. Electron. Ind. Forum*, Boston, MA, May 1997, pp. 43–49.

[30]   Borja, C. and Romeu, J., "Multi-Band Sierpinski Fractal Patch Antenna," *Proc. IEEE Antennas and Propagation Society Int. Symposium*, Vol. 3, July 2000, pp. 1708–1711.

[31] Romeu, J., et al., "High Directivity Modes in the Koch Island Fractal Patch Antenna," *Proc. IEEE Antennas and Propagation Society Int. Symposium,* Vol. 3, July 2000, pp. 1696–1699.

[32] Yeo, J., and Mittra, R., "Modified Sierpinski Gasket Patch Antenna for Multi-Band Applications," *Proc. IEEE Antennas and Propagation Society Int. Symposium,* Vol. 3, July 2001, pp. 134–137.

[33] Anguera, J., et al., "Bow Tie Microstrip Patch Antenna Based on the Sierpinski Fractal," *Proc. IEEE Antennas and Propagation Society Int. Symposium,* Vol. 3, July 2001, pp. 162–165.

[34] Kim, I., et al., "The Koch Island Fractal Microstrip Patch Antenna," *Proc. IEEE Antennas and Propagation Society Int. Symposium,* Vol. 2, July 2001, pp. 736–739.

[35] Gianvittorio, J., and Rahmat-Samii, Y., "Fractal Patch Antennas: Miniaturizing Resonant Patches," *Proc. IEEE Antennas and Propagation Society Int. Symposium and USNCNRSI National Radio Science Meeting MRSI,* July 2001, p. 298.

[36] Gianvittorio, J., "Fractal Antennas: Design, Characterization, and Application," M.Sc. Thesis, University of California, Los Angeles, CA, 2000.

[37] Krzysztofik, W. J., and Baraski, M., "Fractal Structures in the Antenna Applications," *KKRRiT-2007,* Proc. *National Conference on Communication, Radiobroadcasting and Television,* June 2007, Gdansk, Poland, pp. 381–384.

[38] Puente, C., Romeu, et al., "Fractal Multi-Band Antenna Based on the Sierpinski Gasket," *Electronics Letters,* Vol. 32, No. 1, January 1996, pp. 1–2.

[39] Baranski, M., "Multi-Band Fractal Antennas," M.Sc. Thesis, Wroclaw University of Technology, Wroclaw, Poland, 2007.

[40] Rahim, M. K. A., Jaafar, A. S., and Aziz, M.Z.A.A., "Sierpinski Gasket Monopole Antenna Design," *Proc. Asia-Pacific Conference on Applied Electromagnetic,* December 2005, pp. 49–52.

[41] Krzysztofik, W. J., "Fractal Monopole Antenna for Dual-ISM-Bands Applications," *Proc. 36th European Microwave Conference,* September 2006, pp. 1461–1464.

[42] Krzysztofik, W. J.,"Modified Sierpinski Fractal Monopole for ISM-Band Handset Applications," *IEEE Transactions on Antennas and Propagation,* 2008.

[43] Castany, J. S., Robert, J. R., and Puente, C., "Mod-P Sierpinski Fractal Multi-Band Antenna," *Proc. Millennium Conference on Antennas and Propagation,* April 2000.

[44] Song, C. T. P., Hall, P. S., and Ghafouri-Shiraz, H., "Shorted Fractal Sierpinski Monopole Antenna," *IEEE Transactions on Antennas and Propagation,* Vol. 52, No. 10, October 2004, pp. 2564–2570.

[45] Puente, C., "Fractal Antennas," Ph.D. Dissertation, Department of Signal Theory and Communications, Universitat Politecnica de Catalunya, Spain, June, 1997.

[46] Song, C. T. P., et al., "Sierpinski Monopole Antenna with Controlled Band Spacing and Input Impedance," *Electronics Letters,* Vol. 35, No. 13, June 1999, pp. 1036–1037.

[47] Anguera, J., "Fractal and Broadband Techniques on Miniature, Multifrequency, and High-Directivity Microstrip Patch Antennas," Ph.D. dissertation, Department of Signal Theory and Communications, Universitat Politecnica de Catalunya, Spain, 2003.

[48] Gala, D., et al., "Miniature Microstrip Patch Antenna Loaded with a Space-Filling Transmission Line Based on the Fractal Hilbert Curve," *Microwave and Optical Technology Letters,* Vol. 38, No. 4, August 2003, pp. 311–312.

[49]   Azad, M. Z., and Ali, M., "A Miniaturized Hilbert PIFA for Dual-Band Mobile Wireless Applications," *IEEE Antennas and Wireless Propagation Letters,* Vol. 4, 2005, pp. 59–62.

[50]   Guterman, J., Moreira, A. A., and Peixeiro, C., "Microstrip Fractal Antennas for Multistandard Terminals," *IEEE Antennas and Wireless Propagation Letters,* Vol. 3, 2004, pp. 351–354.

[51]   Ghali, H., Moselhy, T., "Design of Fractal Slot Antennas," *Proc. 34th European Microwave Conference,* October 2004, pp. 1265–1268.

[52]   El-Damak, A. R., Ghali, H., and Ragaie, H. F., "Circularly Polarized Fractal Slot Antenna," *Proc. 35th IEEE European Microwave Conference,* Vol. 1, October 2005, pp. 1–4.

[53]   Dehkhoda, P., and Tavakoli, A., "Circularly Polarized Microstrip Fractal Antennas," *Proc. IEEE Antennas and Propagation Society Int. Symposium,* Vol. 4, June 2004, pp. 3453–3456.

[54]   Kim, Y., and Jaggard, D. L., "The Fractal Random Array," *Proc. IEEE,* Vol. 74, No. 9, 1986, pp. 1278–1280.

[55]   Lakhtakia, A., Varadan, V. K., and Varadan, V. V., "Time-Harmonic and Time-Dependent Radiation by Bifractal Dipole Arrays," *International Journal of Electronics,* Vol. 63, No. 6, 1987, pp. 819–824.

[56]   Werner, D. H., and Werner, P. L., "Fractal Radiation Pattern Synthesis," *Proc. USNC/URSI National Radio Science Meeting,* January 1992, p. 66.

[57]   Werner, D. H., Haupt, R. L., and Werner, P. L., "Fractal Antenna Engineering: The Theory and Design of Fractal Antenna Arrays," *IEEE Antennas and Propagation,* Vol. 41, No. 5, October 1999, pp. 37–58.

[58]   Walker, G. J., and James, J. R., "Fractal Volume Antennas," *Electronics Letters,* Vol. 34, No. 16, August 1998, pp. 1536–1537.

[59]   Anguera, I., et al., "Miniature Wideband Stacked Microstrip Patch Antenna Based on the Sierpinski Fractal Geometry," *Proc. IEEE Antennas and Propagation Society Int. Symposium,* Vol. 3, July 2000, pp. 1700–1703.

CHAPTER 5

Miniaturized Integrated Multiband Antennas

Marta Martínez-Vázquez, Eva Antonino-Daviú, and Marta Fabedo-Cabrés

5.1 Design Considerations for Integrated Multiband Handset Antennas

In the general case, antennas are designed and measured in well-defined environments, over either infinite or large ground planes. However, antennas for handheld communications terminals have to operate on an electrically small device, which will cause a distortion of its radiation characteristics. It is thus important to determine how the fact of integrating an antenna in a terminal will affect its actual behavior regarding both input and radiation characteristics [1, 2]. The characteristics of small internal antennas, such as printed inverted-F antennas (PIFAs) mounted on a handset will depend on both the antenna position on the terminal chassis and the dimensions of the chassis (the length in particular). While the typical bandwidth of a patch type antenna on an infinite ground plane lies between 1% and 3%, more than 10% impedance bandwidth is achieved in standard size terminals.

In general, designing a handset antenna is not an easy task, as this type of antenna is subject to very stringent specifications [3]. Small size, light weight, compact structure, low profile, robustness, and flexibility are the prime considerations conventionally taken into account in small antenna design [4]. In addition, as new mobile handsets are required to operate at multiple standards, their antennas are expected to grab as much spectrum as possible, so they may provide multiband or broadband operation [5].

Although the progression in microelectronics has resulted in striking reductions in the size of the terminal, the size of the antenna cannot be shrunk so easily, as it is determined not only by technological, but mostly by physical factors. Extensive research has been carried out in the last two decades to find ways of reducing the size of resonant antennas so that they will fit within a given volume inside a handset [6]. This implies that either the antenna has to be electrically small or conformed to a certain shape adapted to a small volume available [7]. However, this gives rise to restrictions regarding polarization, radiation efficiency, and bandwidth, and furthermore increases the sensitivity to manufacturing tolerances. In the past, some compromises were adopted regarding the performance of the antenna, and certain degradation in its matching and efficiency characteristics was accepted, if the aims of integration and cost were accepted. Yet, the pressure of network operators is changing the trend, and now we can witness new efforts to obtain terminal with more efficient radiating elements.

It is easy to find examples of handset antennas in the literature, so in this chapter we will not deal directly with the antenna element. Extended reviews can be found in [5, 8]. Here, we will only consider a generic handset, such as a bar phone, with an integrated, dual-band patch antenna. The results can be extended to other terminal geometries and antenna types.

5.1.1 Hearing Aid Compatible Handsets

A key challenge for the mobile industry is reducing the interference between some hearing aids and some digital wireless phones. Many hearing aid wearers still experience interference when using mobile phones, particularly if they use a tele-coil for telephone communication. A telecoil is an induction coil that coil converts magnetic energy of a varying magnetic field to electrical energy, retaining the orig-inal information contained in the magnetic field around a mobile phone.

The presence of noise affects the speech comprehension of people with partial hearing much more adversely than that of people with normal hearing. People with partial hearing thus require a high signal-to-noise ratio to optimize their speech discrimination. Telecoils help people with hearing loss achieve this high signal-to-noise ratio by coupling the hearing aid (HA) or cochlear implant (CI) inductively with hearing-aid compatible (HAC) phones.

The radio waves emitted by the cell phone are referred to as radio-frequency (RF) emissions. The RF emissions create an electromagnetic (EM) field with a pulsing pattern around the mobile phone. This pulsing energy may potentially be detected by the microphone of the hearing aid or the telecoil circuitry, and per-ceived by the user as a buzzing sound. Besides, telecoil users may experience base-band, magnetic interference caused by the electronics in the mobile phone (backlighting, display, keypad, battery, circuit board, etc.). This baseband electro-magnetic interference occurs in addition to the RF interference, thus potentially increasing the effect perceived by the user.

In the United States, the Federal Communications Commission (FCC) requires that wireless phone manufacturers and wireless phone service providers make dig-ital wireless phones accessible to individuals who use hearing aids. The timetable provided in June 2005 by the FCC [9] states that all wireless operators should ensure that 50% of their handset models would be hearing aid compatible by Feb-ruary 2008. Yet, this has been recently amended, to reflect new requirements and allow new deadlines for manufacturers and service providers [9abis].

This new HAC requirement translates into new problems related to antenna design. Indeed, the design of HAC terminals will have to be adapted to minimize the interaction with hearing aids. This will include not only the antenna, but also any element that could eventually interfere with a hearing device (loudspeaker, battery, display). Electromagnetic compatibility (EMC) and near-field effects will also have to be considered, whereas measurement setups should also be defined to determine the behavior of such devices.

The design of the telephone handset may be important for hearing aid users. HAC phones have often either a clamshell or flip-up design, where the only part of the phone in the section that flips up is the speaker. This design provides some physical distance between the hearing aid and the components that may potentially cause interference. For telecoil users, it also provides physical distance between the cell phone electronics (another potential source of interference) and the hearing aid. The greater the distance between the hearing aid and these electronics, the less potential there is for interference experienced by the hearing aid wearer.

5.2 Effect of the Terminal on the Antenna Behavior

Due to space and design constraints, all the components inside a handset are tightly packet, leading to strong mutual coupling and interaction. Even the geometry of the antenna itself is determined by criteria that are not related with electromagnetics, such as the available contact points, or the presence of the acoustic devices and/or a camera. Here we will discuss the effect of the integration of a patch antenna into the handset, with special emphasis on the size of the chassis, and the position of the battery and the display.

5.2.1 Effect of the Size of the Chassis

As stated before, the effect of the length of a portable terminal on the bandwidths of PIFA antennas in the 900-MHz band was first reported in [1] and further analyzed in [10]. Indeed, when a terminal antenna is in operation, surface currents will be induced on other conducting elements, especially the PCB, which acts as a ground plane [11]. The operation will thus be similar to that of a folded, asymmetric dipole, as illustrated in Figure 5.1 for a PIFA antenna. In most cases, the antenna is a resonant element, whereas the length of the terminal lies usually between $\lambda_0/4$ and $\lambda_0/2$ at the operating frequency. This has a significant influence upon the behavior of the antenna, regarding its matching, the impedance bandwidth, the radiation patterns, and the interaction with the user.

The operation of the whole system can be modeled as a combination of series and shunt resonant circuits, as first described by Vainikainen et al. in [12] for a single-band terminal antenna. The terminal is thus described as system composed of two coupled resonators: the antenna, which exhibits a quasi-TEM transmission wave mode with high Q, and the printed circuit board (PCB) that represents the chassis of the phone and supports a low-Q current distribution similar to that of a thick wire dipole. This model was later refined in [13] and corrected by Boyle and Ligthart using the radiating and balanced mode theory [14].

Therefore, if the considered antenna is a radiating patch, such as a PIFA, a more realistic simulation model would be the one presented in Figure 5.2. In all the abovementioned models, the resonant frequency of the ground plane is determined by its actual physical size. To illustrate it, let's consider a dual-band patch antenna over a metallic ground plane of dimensions 40 mm × l mm, which is a standard size for bar-type handsets. The structure is described in Figure 5.2. The antenna is tuned to operate in the GSM and DCS frequency bands.

The effect of the ground plane length on the resonance frequency and the bandwidth was analyzed by varying the length l. Figure 5.3 shows how the resonance frequency changes depending on the length l of the PCB. This variation is given in function of the center frequency of each band of interest. The relative resonance frequency f_{rel} is given by $f_{rel}=f_r/f_0$, with f_r being the achieved resonant frequency, and f_0 the center frequency of the band of interest (920 MHz in the GSM band, and 1,975 MHz in the DCS band)

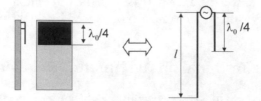

Figure 5.1   Dipole equivalent of a PIFA over a finite ground plane.

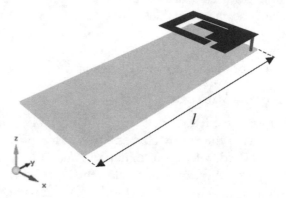

Figure 5.2   Simulation model of a dual-band patch antenna over a metallic ground.

The evolution is different in the two bands of interest. Indeed, the length of the ground is close to $\lambda/4$ in the case of the GSM 900, but equals $\lambda/2$ in the GSM 1800 band. The ground plane contributes then to the total radiation of the structure, by acting as a counter pole to the actual antenna [11, 15].

Not only the resonance frequency, but also the bandwidth is affected by the length of the ground plane. In some cases, variations of up to 300% have been reported by tuning dimensions of the terminal [16]. In [1] the optimum length of a terminal was situated around $0.4\lambda_0$, with λ_0 being the wavelength at the desired operating frequency. The results presented in [17, 18] confirm that conclusion. Indeed, when the length of the PCB approaches $0.4\lambda_0$, its effective length is close to $\lambda_0/2$, which is the resonant length of a dipole. In fact, for thick dipoles, the resonance occurs when its length is slightly shorter than $\lambda_0/2$.

Figure 5.4 shows the evolution of the 6-dB relative bandwidth in the case of the structure described above. Again, the effect seems to be different for each of the two frequency bands, as indicated in Figure 5.4(a). But if the variation is studied in function of the wavelength at the frequencies of interest, more similar behavior can be observed, as displayed in Figure 5.4(b). The expected bandwidth maximum appears in both cases for a PCB length between $0.3\ \lambda_0$ and $0.4\lambda_0$. In the case of DCS, a second maximum can be observed when around $0.8\lambda_0$. This would also be the case for the GSM band if a larger PCB had been considered [17, 19].

In fact, the bandwidth of the system antenna plus chassis is not determined by the antenna, but mainly by the length of the PCB. The bandwidth maximum is achieved when both the antenna and the ground plane reach the resonance state simultaneously. For a dual-band terminal, the ideal size would be the one that allows for a good compromise between both bands. Yet, in practical applications,

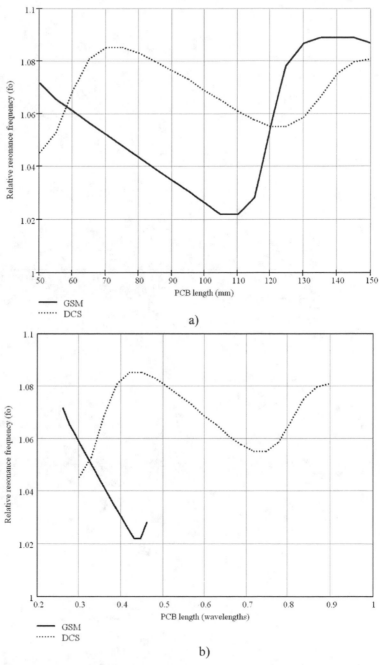

Figure 5.3   Relative variation of the resonance frequency for different terminal lengths in (a) millimeters and (b) wavelengths.

the size of the ground plane is determined by aspects that are not related to its electromagnetic behavior.

The bandwidth criterion is a key point in the design of an antenna for a terminal. In some cases, the desired bandwidth cannot be obtained with the given ground plane dimensions. It is thus necessary to find ways to artificially change the electric length of the structure. One way of doing it would be cutting slots in

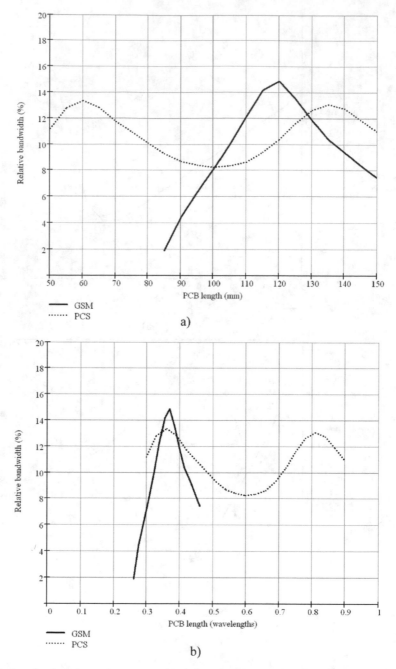

Figure 5.4 Evolution of the achieved relative –6-dB impedance bandwidth versus terminal length in the GSM 900 and 1,800 bands versus length of the terminal in (a) millimeters and (b) wavelengths.

the ground plane as first suggested in [13]. There, a slot was used to lengthen electrical size of the PCB in the 900-MHz band, and thus lower the resonance frequency of the PCB. This idea has been further developed [20–23] and used in industrial applications [24, 25]. As the currents on the PCB are concentrated mainly in its edges [18], a new strategy has been recently proposed to optimize the bandwidth of dual-band handset terminals. It consists of introducing wave traps along the sides of the chassis to shorten it [26].

The effect of the PCB width and thickness has also been studied by various authors [18, 19, 27]. For the antenna structure considered here, the effect of varying the PCB width is depicted in Figures 5.5 and 5.6. If the PCB is narrower than the patch antenna, the bandwidth can be considerably increased in some cases, if the coupling to the ground plane becomes more effective [13]. Some of these studies show also that the PCB thickness has almost no effect on the characteristics of the system.

Section 5.3.2 will show some examples of how to use the radiating properties of the ground plane in a constructive way, to design radiating systems using the theory of characteristic modes.

5.2.2   Effect of the Handset Components

As already mentioned, not only the size of the terminal, but also the presence of other elements will have a strong influence in the matching, impedance bandwidth, and radiation behavior of a handset antenna. All the metallic and dielectric components that are needed to implement a mobile phone will interact and affect the performance of the system, and should thus be taken into account from the beginning of the design phase. Figure 5.7 shows an example of simulation model for a typical mobile phone terminal, without its external cover.

Some examples of these components are:

- The antenna radiation;
- The battery;
- The vibration motor;
- The acoustic components;
- The display: the currents induced onto this component can contribute to an increase of the specific absorption rate (SAR):
- The casing, which is made of plastic material and will increase the losses and detune the resonant frequency of the antenna;
- The RF shielding of the circuits in the terminal;
- Battery connector, I/O connectors, memory card readers.

Some of those components, like RF shield or acoustic devices, may even be mounted under the antenna, thus changing its volume and modifying its behavior. In other cases, the geometry of the antenna will have to be adapted to allow the integration of certain element, like a camera. In the case of the acoustic components, difficulties are added due to the fact that they include magnetic materials, and the fabrication tolerances can be relatively high. Their connection to the ground introduces a capacitive effect that distorts the performance of the antenna. In the next paragraphs we will consider some examples that illustrate the effect of the battery and the display.

5.2.2.1   Influence of the Battery

One of the most significant distortions for the antenna behavior is caused by the coupling to the battery casing. Indeed, the presence of this element and its posi-

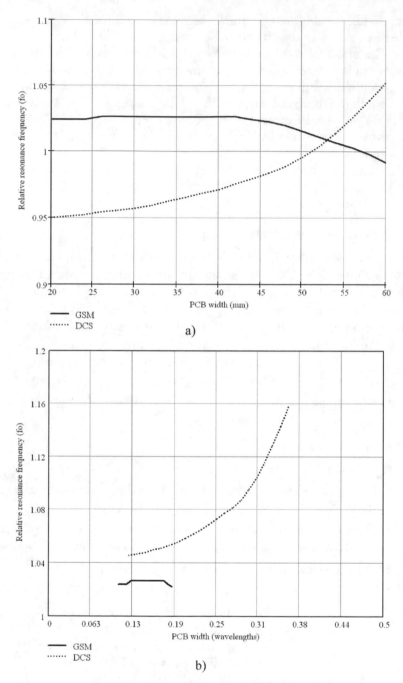

Figure 5.5   Relative variation of the resonance frequency for different terminal lengths in (a) millimeters and (b) wavelengths.

tion on the PCB will change the current distribution, and thus the performance of the whole system.

To illustrate the effect of the battery, we will consider the same antenna as in Section 5.1. The battery is modeled as a 50 mm × 36 mm × 7 mm metallic box, at a distance d from the antenna patch, as displayed in Figure 5.8. The ground plane size was fixed to 100 mm × 40 mm. The obtained results regarding the variation

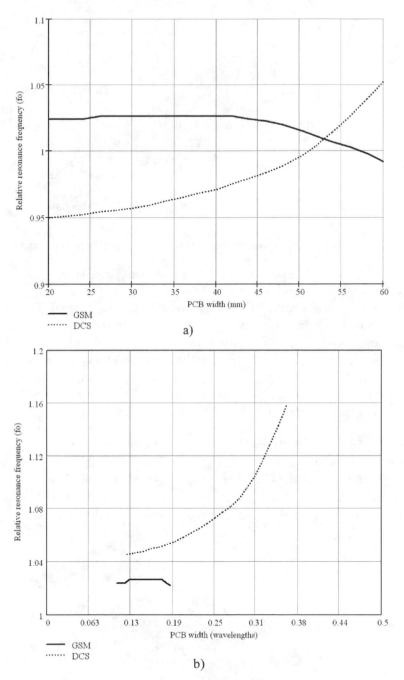

Figure 5.6   Evolution of the achieved relative –6-dB bandwidth versus terminal width in the GSM 900 and 1,800 bands versus length of the terminal in (a) millimeters and in (b) wavelengths.

of the resonance frequency and the 6-dB relative impedance bandwidth are presented in Figures 5.9 and 5.10, respectively. In both cases, it can be appreciated how the effect is more significant in the GSM band than in the DCS band. In any case, this effect is not so significant compared to that of the chassis length.

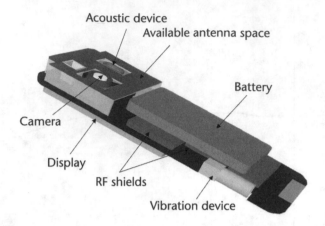

Figure 5.7   Example of simulation model of a mobile communications handset (bar phone).

Recently a similar study on the battery effects on a small handset PIFA in the UMTS frequency band has been presented [28].

A more practical example is presented in Figures 5.11 and 5.12. In this case, a realistic handset prototype was built and measured [29]. The system combines a PIFA antenna with a slot in the ground plane to provide coverage in the GSM, DCS, PCS, UMTS, and SDMB (Satellite Digital Media Broadcast used in Korea) frequency bands [24, 25, 30, 31]. Different parameters are analyzed, namely antenna efficiency, radiation efficiency, and input return loss (S_{11}) in different cases: without battery and with three different battery positions.

5.2.2.2   Effect of the Display

Another element that has a significant influence on the radiation performance is the display, especially if slot antennas or slotted ground planes are considered. Handsets feature ever larger displays, designed for multimedia applications. While this may have the effect of enlarging the width of the terminal and thus the vol-

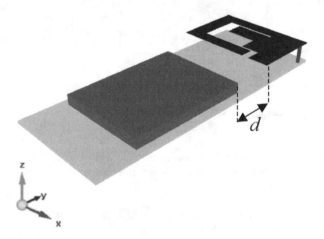

Figure 5.8   Simulation model of the antenna over a PCB with a battery.

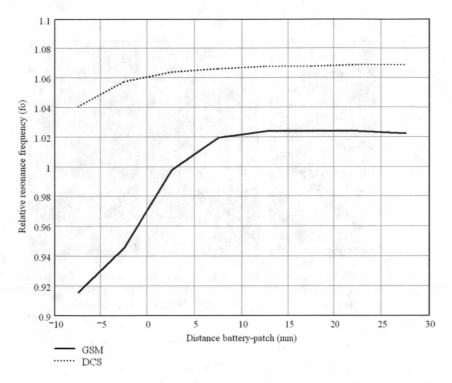

Figure 5.9 Variation of the relative resonance frequency versus distance antenna patch battery.

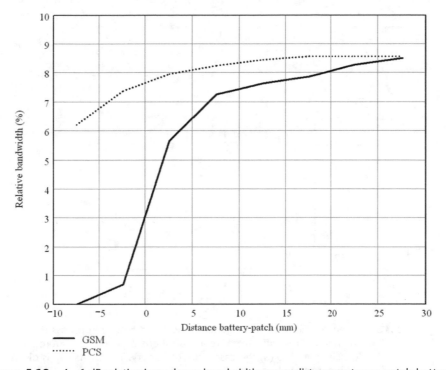

Figure 5.10 A −6-dB relative impedance bandwidth versus distance antenna patch battery.

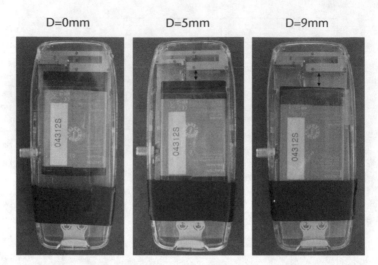

Figure 5.11 Handset prototype with different distances antenna battery. (Courtesy of Fractus S.A. The design is protected by one or more of the following patents and patents applications: EP1223637, WO 2001/22528, WO 2006/070017, WO 2003/023906.)

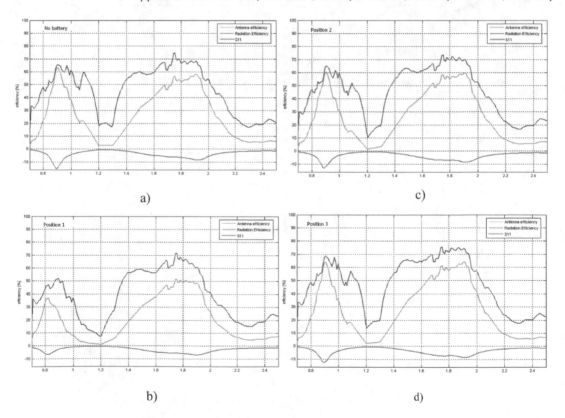

Figure 5.12 Antenna efficiency, radiation efficiency, and input return loss (S_{11}) of the handset antenna for the different antenna positions described in Figure 5.11 (b), (c), and (d), compared to (a) the case without battery. The frequency is given in gigahertz, the input return loss in decibels.

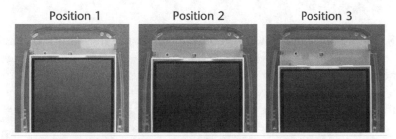

Figure 5.13   Handset prototype with different positions of the display. (Courtesy of Fractus S.A. The design is protected by one or more of the following patents and patents applications: EP1223637, WO 2001/22528, WO 2006/070017, WO 2003/023906.)

ume available to integrate the antenna, it can also heighten the influence of the display on several characteristics, especially concerning SAR levels.

As an example, the effect of the position of the display on the same realistic terminal as in the section "Influence of the Battery" is presented in Figures 5.13 and 5.14. In this case, the effect of the display is quite significant, as its position with respect to the slot in the ground plane has a major influence on the impedance matching, bandwidth and efficiency [29].

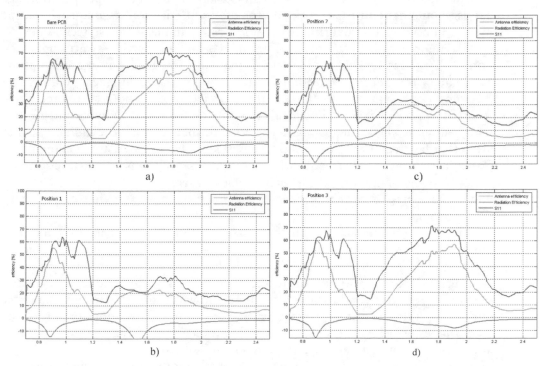

Figure 5.14   Antenna efficiency, radiation efficiency, and input return loss (S11) of the handset antenna for the different display positions described in Figure 5.13 (b), (c), and (d), compared to the case without display (a). The frequency is given in gigahertz, the input return loss in decibels.

5.3 Modal Analysis

In Section 5.1, some practical considerations for the design of handset antennas were addressed. These design hints were related to the effect of the terminal components over the antenna, rather than to the antenna topology itself. Due to its practical nature, a total control over these effects can only be obtained through years of investigation on mobile terminal characterization.

Unfortunately, as the antenna geometry complicates, more often than not, there is no closed formulation to analyze it and the use of numerical methods [32, 33] becomes imperative. In consequence, design of modern handset antennas relies on the use of self-developed numerical codes or commercial electromagnetic simulators, such as IE3D, FEKO, EMPIRE, or HFSS, among others, to evaluate antenna performance before a physical prototype is fabricated. Under this circumstance, time for antenna design can be dramatically reduced using computers. In any case, even with the support of computers, the success of the final design depends upon the intuition and previous experience of the designer, and in most cases, the final optimization is in fact made by "cut and try" methods. As a result, these days antenna design is very much governed by designer expertise and know-how.

On the other hand, an alternative and certainly in vogue approach to design handset antennas consists of using automated optimization techniques based on pseudorandom search algorithms [34]. Typical examples of these techniques are genetic algorithms [35], artificial neural networks [36], particle swarm optimization [37], or bees' algorithms [38]. Their main advantage is that once the optimization algorithm is programmed, little interaction with the designer is required, as the computer is supposed to arrive at the expected specifications autonomously.

As a matter of fact, although all the above-mentioned design strategies are really suitable when time to market is critical, their major problem is that they are rather lacking in physical insight, so real knowledge of the antenna operating principles is mislaid. Consequently, publications giving useful instructions for better antenna design are scarce. Nevertheless, there exist other not-so-common design strategies such as the theory of characteristic modes, that can alleviate this problem.

The theory of characteristic modes, first developed by Garbacz [39] and later refined by Harrington and Mautz in the seventies [40, 41], was originally applied to antenna shape synthesis [42, 43], and control of obstacle scattering by reactive loading [44]. It was also applied to the analysis of slots in conducting cylinders [45] or in perfectly conducting planes [46]. However, this theory practically fell into disuse later, in spite of the fact that it leads to modal solutions even for arbitrary shapes, which is particularly useful in problems involving analysis, synthesis and optimization of antennas and scatterers [47–49].

By definition, characteristic modes are current modes obtained numerically for arbitrarily shaped conducting bodies. These modes present really appealing properties as not only do they make possible a modal analysis of conducting objects, but they also bring valuable information for antenna design, as they provide physical interpretation of the radiation phenomena taking place on the antenna.

Since characteristic modes are independent of any kind of excitation, they only depend on the shape and size of the conducting object. Thus, antenna design using characteristic modes can be performed in two steps. First, the shape and size of the radiating element are optimized. If the size of the element is scaled, the resonant frequency of modes will be modified, whereas if the shape of the element is varied, not only the resonant frequency, but also the radiating properties of modes will change. Next, the optimum feeding configuration is chosen so that the desired modes may be excited. For modeling electrically small conducting bodies few modes are needed. Thus, small and intermediate size antennas can be fully characterized in a wide operating band just considering three or four characteristic modes.

The study presented next is intended to illustrate that characteristic modes can be effectively used to carry out a controlled design of handset antennas. In the following sections, characteristic modes are used to perform systematic analysis and design of different types of planar antennas suitable for handsets. Results are obtained using a method of moment code based on the mixed potential integral equation (MPIE) [50] and Rao-Wilton-Glisson (RWG) basis functions [51]. This code has been expressly developed to compute characteristic modes efficiently at a wide frequency band.

Although the theory of characteristic modes is extensively described in [40] and [41], the next section includes, for the sake of completeness, a revision of the mathematical formulation of this theory. For illustration purposes, numerical examples for a very well-known structure, such as a rectangular plate, are presented.

5.3.1 The Theory of Characteristic Modes

As explained in [40], characteristic modes or characteristic currents can be obtained as the eigenfunctions of the following particular weighted eigenvalue equation

$$X\left(\vec{J}_n\right) = \lambda_n R\left(\vec{J}_n\right) \tag{5.1}$$

where λ_n are the eigenvalues, $\vec{J}_n$ are the eigenfuncions or eigencurrents, and R and X are the real and imaginary parts of the impedance operator

$$Z = R + jX \tag{5.2}$$

This impedance operator is got after formulating an integro-differential equation. It is known from the reciprocity theorem that if Z is a linear symmetric operator, then its Hermitian parts, R and X, will be real and symmetric operators. From this, it follows that all eigenvalues λ_n in (5.1) are real, and all the eigenfunctions $\vec{J}_n$ can be chosen real or equiphasal over the surface in which they are defined [40]. Moreover, the choice of R as a weight operator in (5.1) is responsible for the orthogonality properties of characteristic modes described in [42], which can be summarized as

$$\left\langle \vec{J}_m^*, R\left(\vec{J}_n\right)\right\rangle = \delta_{mn} \tag{5.3}$$

$$\left\langle \vec{J}_m^*, X\left(\vec{J}_n\right)\right\rangle = \lambda_n \delta_{mn} \tag{5.4}$$

where δ_{mn} is the Kronecker delta (0 if $m \neq n$ and 1 if $m = n$).

Consistent with (5.1), characteristic modes $\vec{J}_n$ can be defined as the real currents on the surface of a conducting body that only depend on its shape and size, and are independent of any specific source or excitation. In practice, to compute characteristic modes of a particular conducting body, (5.1) needs to be reduced to matrix form, as explained in [41], using a Galerkin formulation [50].

$$[X]\vec{J}_n = \lambda_n[R]\vec{J}_n \tag{5.5}$$

Next, eigenvectors $\vec{J}_n$ and eigenvalues λ_n of the object are obtained by solving the generalized eigenproblem of (5.5) with standard algorithms [52].

As an example, Figure 5.15 illustrates the current distribution at first resonance (f =2.4 GHz) for the first six eigenvectors of a rectangular plate of width W=4 cm and length L=6 cm. Computation of these eigenvectors was made using 128 RWG functions for expansion and testing. All currents in Figure 5.15 have been normalized to its maximum value in order to facilitate comparison. Additionally, for a better understanding, Figure 5.16 yields current schematics of these six modes. Eigenvector J_0, as it will be verified later, presents special inductive nature due to its currents forming closed loops over the plate. Eigenvectors J_1 and J_2, which are characterized by horizontal and vertical currents, respectively, are the most frequently used modes in patch antenna applications, while the rest of eigenvectors, J_3, J_4, and J_5, are higher-order modes that might be taken into consideration only at highest frequencies.

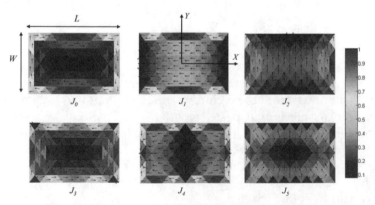

Figure 5.15   Normalized current distribution at first resonance (f =2.4 GHz) of the first six eigenvectors J_n of a rectangular plate of width W=4 cm and length L= 6 cm.

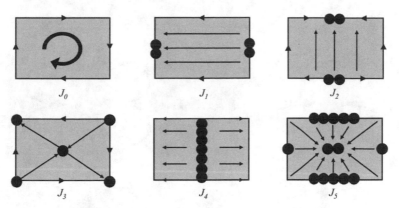

Figure 5.16 Current schematics for the six eigenvectors shown in Figure 5.15.

It is worth noting that eigenvectors presented in Figure 5.15 have been computed in free space; however, the presence of a ground plane below the plate would not alter significantly their current distribution, although it would affect their resonance and radiating bandwidth [53]. Note also that due to eigenvectors dependency upon frequency, if a structure is to be analyzed in a frequency range, modes will need to be recalculated at every frequency [54].

On the other hand, electric fields E_n produced by characteristic currents J_n on the surface of the conducting body are called characteristic fields [40]. From (5.1) it can be derived that these characteristic fields can be written as

$$E_n(\vec{J}_n) = Z(\vec{J}_n) = R(\vec{J}_n) + jX(\vec{J}_n) = R(\vec{J}_n)(1 + j\lambda_n) \tag{5.6}$$

Then, from (5.6), it is extracted that characteristic electric fields are equiphasal, since they are $(1+j\lambda_n)$ times a real quantity. Orthogonality relationships for characteristic electric fields can be reached from characteristic currents by means of the complex Poynting theorem

$$P(J_m, J_n) = \left\langle J_m^*, ZJ_n \right\rangle = \left\langle J_m^*, RJ_n \right\rangle + j\left\langle J_m^*, XJ_n \right\rangle$$
$$= \oiint_{S'} \vec{E}_m \times \vec{H}_n^* ds + j\omega \iiint_{\tau'} \left(\mu \vec{H}_m \cdot \vec{H}_n^* - \varepsilon \vec{E}_m \cdot \vec{E}_n^* \right) d\tau \tag{5.7}$$
$$= (1 + j\lambda_n)\delta_{mn}$$

Figure 5.17 depicts the azimuthal radiation pattern ($\theta = 90°$) at 4 GHz for the modal electric fields E_n produced by the current modes J_n of the rectangular plate. It can be observed that the radiation pattern generated by mode J_0 presents nearly omnidirectional characteristic, while the rest of modes present growing number of lobes as the order of the mode increases.

Due to the above-mentioned orthogonality properties over both the surface of the body and the enclosing sphere at infinity, characteristic modes radiate power independently of one another. Because of this attractive feature, characteristic

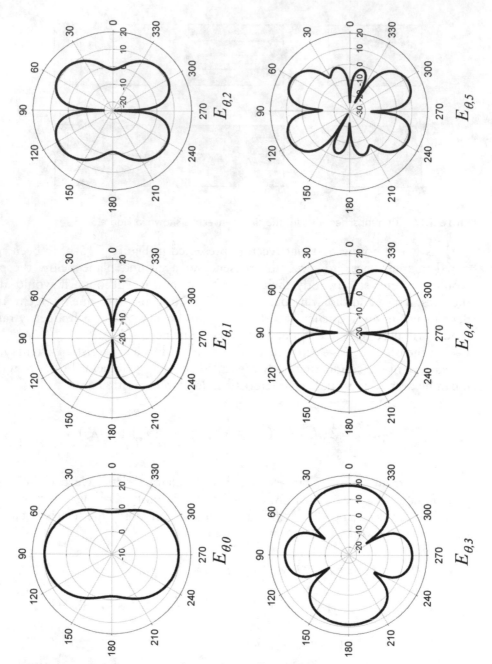

Figure 5.17 Azimuthal radiation pattern ($\theta = 90°$) at 4 GHz of the modal electric fields $E_{\theta,n}$ produced by the current modes J_n of Figure 5.15.

modes can be used as a basis set in which to expand the unknown total current J on the surface of the conducting body as

$$J = \sum_n \frac{V_n^i J_n}{1 + j\lambda_n}$$

(5.8)

The term V_n^i in (5.8) is called the modal excitation coefficient [40], and it is defined as

$$V_n^i = \langle J_n, E^i \rangle = \oint_n J_n \cdot E^i ds \qquad (5.9)$$

The modal excitation coefficient accounts for the way the position, magnitude, and phase of the applied excitation influence the contribution of each mode to the total current J. Consequently, the product $V_n^i J_n$ in (5.8) models the coupling between the excitation and the nth mode, and determines if a particular mode is excited by the antenna feed or incident field.

The term λ_n in (5.8) corresponds with the eigenvalue associated to the nth characteristic mode. This eigenvalue is of utmost importance because its magnitude gives information about how well the associated mode radiates. From the complex power balance in (5.7) it is deduced that power radiated by modes is normalized to unit value. In contrast, reactive power is proportional to the magnitude of the eigenvalue. Considering that a mode is at resonance when its associated eigenvalue is zero ($|\lambda_n = 0|$), it is inferred that the smaller the magnitude of the eigenvalue is, the more efficiently the mode radiates when excited. In addition, the sign of the eigenvalue determines whether the mode contributes to store magnetic energy ($\lambda_n > 0$) or electric energy ($\lambda_n < 0$).

Figure 5.18 shows the variation with frequency of the eigenvalues for the six current modes of the rectangular plate presented before.

It is observed that all eigenvalues begin by being negative, next they resonate ($\lambda_n = 0$), and finally they keep a small constant positive value, except for eigenvalues λ_0 associated to mode J_0, which are positive at every frequency. This means that mode J_0, inherent in planar structures and wire loops, exhibits special behavior as it does not resonate. If this mode were excited, it would only contribute to increase magnetic reactive power. For the particular case of the rectangular plate,

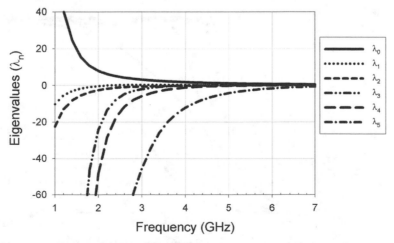

Figure 5.18   Variation with frequency of the eigenvalues λ_n associated to the current modes J_n of the rectangular plate depicted in Figure 5.15.

as eigenvalues continue with very small value after resonance, it is difficult to identify where each curve passes though zero and hence at which frequency each mode resonates.

Finally, it is worth mentioning that the real nature of characteristic modes derived from (5.1) constitutes an advantage in comparison with complex natural modes directly obtained from impedance matrix [Z]. Working with complex basis functions deals to complexity increase in computation, since it is necessary to give different treatment to the real and imaginary parts of the current to get at accurate results [55]. Another drawback of natural modes is that their eigenvalues are also complex, thus they are not so easy to analyze and explain physically. The next section explains in detail how to make the most of the information provided by eigenvalues to perform a complete modal characterization of an antenna.

5.3.1.1 Physical Interpretation of Characteristic Modes

As exposed before, an analysis of the eigenvalue variation with frequency is very useful for antenna design as it brings information about the resonance and the radiating properties of the current modes. Nevertheless, in practice other alternative representations of the eigenvalues are preferred.

Since the modal expansion of the current described in (5.8) is inversely dependent upon eigenvalues, it seems more consistent analyzing the variation of the term $\left|\frac{1}{1+j\lambda_n}\right|$ rather than the variation of the isolated eigenvalue. This term is usually called modal significance (MS), as it represents the normalized amplitude of the current modes [49]. This normalized amplitude only depends on the shape and size of the conducting object, and does not account for excitation. Figure 5.19 depicts the variation with frequency of the modal significance related to current modes J_n of the rectangular plate of Figure 5.15.

The resonance of each mode can be identified by a maximum value of one in the modal significance curves. This means that the nearest the curve is to its maximum value, the most effectively the associated mode contributes to radiation. Then, the radiating bandwidth of a mode can be established according to the

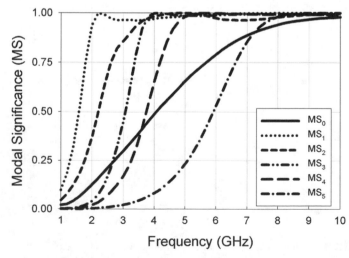

Figure 5.19   Variation with frequency of the modal significance (MS) related to the current modes J_n of Figure 5.15.

width of its modal significance curve near the maximum point. As shown in Figure 5.19, for the case of the rectangular plate of dimension 4 cm × 6 cm, all characteristic modes, except for mode J_0, present quite efficient radiating behavior as their MS curve stands slightly below the maximum value of one after resonance.

However, there exists another even more intuitive representation of the eigenvalues that is based on the use of characteristic angles. Characteristic angles are defined in [56] as

$$\alpha_n = 180° - \tan^{-1}(\lambda_n) \tag{5.10}$$

From a physical point of view, characteristic angle models the phase difference between a characteristic current J_n and the associated characteristic field E_n. Figure 5.20 presents the variation with frequency of the characteristic angle α_n associated to the current modes of the rectangular plate of Figure 5.15.

Observe that modes resonate when $\lambda_n=0$; that is, when its characteristic angle is $\alpha_n=180°$. Therefore, when the characteristic angle is close to 180° the mode is a good radiator, while when characteristic angle is near 90° or 270°, the mode mainly stores energy. Thus, the radiating bandwidth of a mode can be obtained from the slope at 180° of the curve described by characteristic angles.

Although the information given by Figure 5.20 could have also been extracted from Figure 5.18 or Figure 5.19, characteristic angle representation is often preferred as it is the most intuitive one. From Figure 5.20 the resonance frequency of each mode can be easily identified by looking for the points where $\alpha_n=180°$. Hence, mode J_1 resonates at 2.2 GHz, mode J_2 at 5 GHz, mode J_3 at 4.2 GHz, mode J_4 at 5.6 GHz, and mode J_5 at 10 GHz. The special nature of the nonresonant inductive mode J_0 is also observed in Figure 5.20, since its associated angle remains below 180° at every frequency.

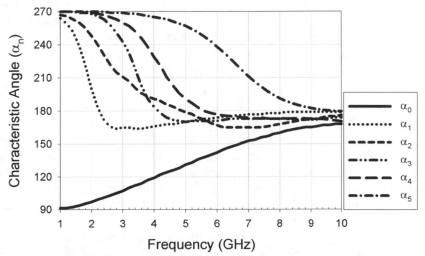

Figure 5.20 Variation with frequency of the characteristic angle α_n associated to the current modes of the rectangular plate in Figure 5.15.

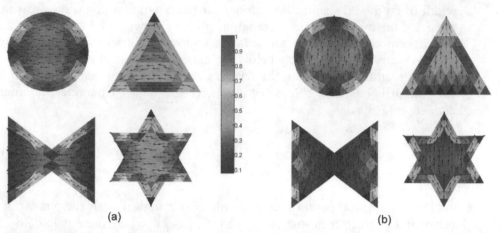

(a)   (b)

Figure 5.21   Normalized current distribution at 2.4 GHz of the first eigenvectors of several planar
                   geometries: (a) horizontal current mode (J_1), and (b) vertical current mode (J_2).

Finally, it should be emphasized that the modal study presented here for the
rectangular plate could have also been performed for planar structures of any
shape. It is worth mentioning that characteristic modes present quite a predictable
behavior in planar structures, whichever their shape is. As a way of example, Figure 5.21 shows the normalized current distribution at 2.4 GHz of the first eigenvector J_1, and the second eigenvector J_2, of several planar geometries. From these
results it can be derived that first eigenvector J_1 is always characterized by horizontal current flow, except for the contour where it follows the perimeter of the
structures. Likewise, the second eigenvector J_2 presents vertical currents along the
different plates with the exception of the contour.

5.3.2 Practical Applications

As explained before, the theory of characteristic modes brings clear insight into
the physical phenomena taking place in a structure, and provides very useful
information to improve its design. For the sake of demonstration, this section is
focused on presenting some handset antenna designs that have been achieved by
direct application of the theory of characteristic modes.

Recently, new design strategies have been explored in order to increase the
radiation efficiency of handset antennas. An example of innovative design is one
that considers the printed circuit board (PCB) of the mobile as part of the antenna
[13]. Since the mobile PCB, which acts as the antenna ground plane, presents resonant dimension at mobile frequencies, its shape and size affect the antenna performance in a significant way. In fact, at lowest frequencies of operation, the PCB
is the main radiator, while the antenna only works as a probe to excite the PCB
current modes. Obviously, to design an antenna from this new perspective, an in-
depth knowledge of the current modes of the structure is needed. To that purpose,
the theory of characteristic modes may be very helpful [57].

Usually, the PCB of a mobile handset consists of a rectangular plate of some
dimensions. From the characteristic mode analysis of the rectangular chassis
made in Section 5.3.1, it can be concluded that horizontal current modes will be

in principle more suitable to be excited in the PCB in order to obtain optimum radiation behavior.

Once the resonant modes of the handset PCB are extracted and its properties identified, two different strategies can be followed in order to excite the more interesting ones: On one hand, the whole geometry of the antenna-chassis structure can be first optimized, and then a wideband feeding mechanism can be arranged to excite those modes with more interesting properties. On the other hand, another strategy can consist in using a compact nonresonant coupling element to optimally couple the energy to the chassis modes, and designing a matching circuit to match the impedance of the coupling element-chassis combination at resonance. Both design strategies will be discussed next.

5.3.2.1 Wideband Feeding of Mobile Chassis

In order to excite the longitudinal current wavemodes in the chassis, a radiating structure consisting of a folded chassis can be proposed [57] Figure 5.22 shows the geometry of the antenna, which resembles a PIFA with a shorting wall. The dimensions of the structure are the following: L=100 mm; W=40 mm; L_{sup}=49.15 mm; h=10 mm; L_{feed}=72.5 mm; h_0=0.5 mm; W_{short}=35 mm; W_{feed}=25 mm. The loop effect of the geometry and the selected dimensions facilitate the excitation of longitudinal modes in the structure, as far as an adequate feeding configuration is selected.

It is worth noting that the folding of the chassis implies a modification of the geometry of the antenna and therefore the properties of the characteristic modes of the chassis are modified, compared with those of the isolated chassis. Although the modal current distributions of the folded chassis are similar to those of the isolated chassis, the characteristic angle associated to the modes changes, especially for the first longitudinal mode.

Figure 5.23 shows the normalized current distribution at first resonance (1.25 GHz) for the first six characteristic modes of the folded chassis. Arrows have been plotted together with characteristic currents for a better understanding of the current flow.

As depicted in Figure 5.23, modal current distributions are very similar to those of the rectangular chassis analyzed in Section 5.3.1. As observed in Figure

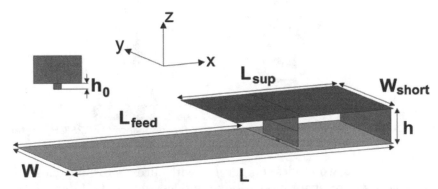

Figure 5.22 Geometry of the proposed folded PCB antenna, excited with a square planar plate. A detail of the feeding plate is provided in the figure.

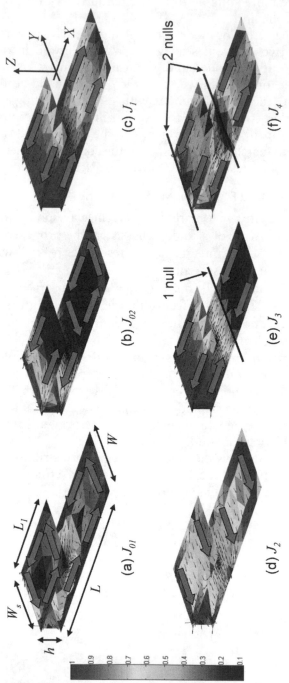

Figure 5.23 Normalized current distribution at first resonance (f =1.25 GHz) of the first six characteristic modes J_n of a folded radiating ground plane.

5.23, there are two modes, J_{01} and J_{02}, with currents forming closed loops, which are special nonresonant modes. Other modes, such as J_1, J_3, and J_4, exhibit longitudinal currents along the structure. Mode J_1 is the fundamental mode, and it flows uninterrupted from the open end on the upper plate to the open end on the bottom plate. This fundamental mode resonates when the current path is approx-

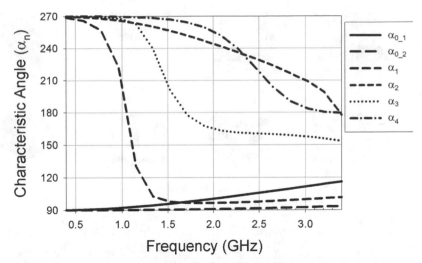

Figure 5.24 Characteristic angle variation with frequency for the first six characteristic modes of the folded radiating chassis.

imately half a wavelength, and it is a folded version of mode J_1 in Figure 5.15. Modes J_3 and J_4, are higher-order longitudinal modes that present one current null and two current nulls along the structure, respectively. Finally, mode J_2 is the only one that presents transverse currents.

The resonance frequency and radiating bandwidth of the above described current modes can be obtained from characteristic angles. Figure 5.24 plots characteristic angles associated to the current modes in Figure 5.23 as a function of frequency. As observed, the properties of characteristic angles change when comparing to those of the isolated chassis shown in Figure 5.20.

As in the rectangular chassis, modes J_{01} and J_{02} do not resonate and present inductive contribution at every frequency. Longitudinal current modes, $J_1, J_3,$ and J_4 resonate at 1.25, 1.7, and 3.25 GHz, respectively, while the transverse mode J_2 resonates at 3.5 GHz. Note that modes J_1 and J_3 resonate at a lower frequency than in the isolated chassis, due to the effect of the capacitive coupling when folding the chassis. Wide bandwidth is still obtained for mode J_3, in contrast to mode J_1 whose radiation behavior has become poorer, since it exhibits a characteristic angle curve with steep slope near 180°. Modes J_2 and J_4 are also good radiators and they resonate at higher frequencies. As observed, currents in the parallel plate region flow with opposite phase at the lower and upper plates for modes J_1 and J_3. This means that there exists some cancellation between these currents flowing like in a transmission line. Conversely, the longitudinal current mode J_4 presents the broadest radiating bandwidth because its currents flow in phase in the parallel plate region, so its radiation reinforces.

Once the antenna-chassis geometry has been selected and the modal analysis has been performed, the next step is to select an optimum feeding arrangement to properly excite the desired modes. For the case of a handset antenna for a cellular phone, longitudinal current modes seem to be the most convenient modes to excite, as they resonate close to GSM900, DCS1800, PCS1900, and UMTS operating bands. At the same time these modes present reasonably good radiating

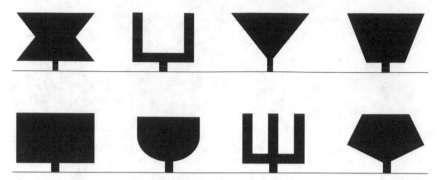

Figure 5.25   Different types of wideband feeding mechanism. The bottom lines represent ground plane connections.

bandwidth, in special mode J_3, which is supposed to provide a matched band wide enough to cover DCS1800, PCS1900, and UMTS services.

The optimum feeding configuration should produce a voltage difference in the structure so as to favor the appearance of the current distribution associated to longitudinal modes. Moreover, it should also present wide matching bandwidth to be able to make the most of the behavior of mode J_1 and especially of the wideband mode J_3. Some references can be found in the literature, which investigate the configuration of wideband feeding mechanisms [56, 58–60]. They consist basically of a planar plate with different shapes, whose dimensions can be adjusted for wideband matching. Figure 5.25 shows some examples of wideband feeding mechanisms for planar antennas.

For the folded chassis, the use of a conventional coaxial probe feeding should therefore be discarded, since it yields narrowband matching and degrades the current distribution of longitudinal modes through the increase of the transverse current component. Figure 5.26 shows a comparison of the return loss for the folded chassis fed with a coaxial probe and with a wideband feeding consisting of a square plate, whose dimensions have been optimized to achieve optimum performance.

As observed, the rectangular plate provides a sort of distributed voltage over the upper part of the folded chassis, which reinforces the excitation of longitudinal modes and avoids transversal modes that can degrade the response of the antenna. Thus, by means of this wideband feeding a much wider impedance bandwidth can be achieved for the structure. Considering a reference value of –6 dB, the impedance bandwidth of this antenna extends from 1 to 6 GHz. Moreover, omnidirectional radiation behavior, very high radiation efficiency, and good gain levels are obtained within this operating bandwidth, as discussed in [57].

As already mentioned in Section 5.2.1, some authors have proposed the insertion of slits in the PCB, in order to further improve the impedance matching level [20, 61]. This has been done in the lower plate of the previous structure as shown in Figure 5.27, with L_f =72.5 mm, R_1=48 mm, and R_2=65.25 mm. In this case, the shape of the feeding plate is a bow tie, which can be adjusted according to a higher number of parameters allowing for an easier optimum matching. The slits, of width 2 mm and length 25 mm, not only produce a meandering effect that reduces the resonant frequencies of longitudinal modes, but also change the current distribution of these modes close to the source, favoring their excitation. This

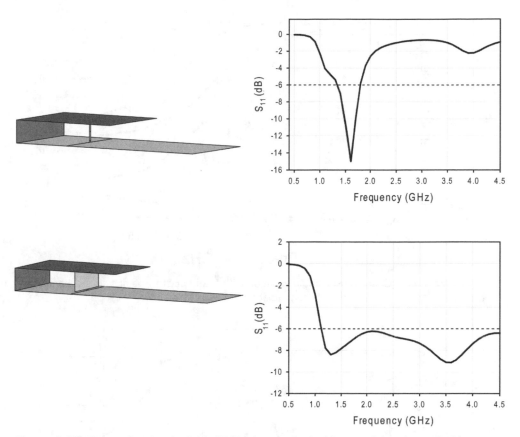

Figure 5.26 Return loss for the folded PCB when excited with a coaxial probe and with a rectangular plate.

reduction in the resonance frequency of modes is confirmed by Figure 5.28, which presents the variation with frequency of the characteristic angles associated to the first seven modes of the slotted folded ground plane. Mode J_5, which has not been represented before, is a higher order longitudinal mode that presents three current nulls along the structure.

Figure 5.29 analyzes the contribution of each mode to the total power radiated by the antenna. Note that the first power maximum, approximately at 0.9 GHz, is caused by mode J_1, the second maximum at 1.8 GHz is due to the excitation of mode J_3, and the third maximum at 3.2 GHz results from the contribution

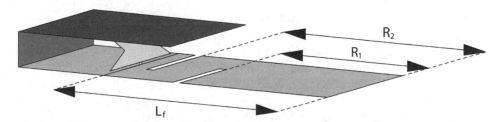

Figure 5.27 Geometry of the folded radiating ground plane with bow-tie-shaped feeding plate.

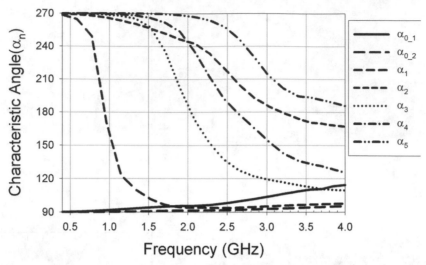

Figure 5.28 Characteristic angle variation with frequency for the first seven characteristic modes of the folded slotted radiating ground plane.

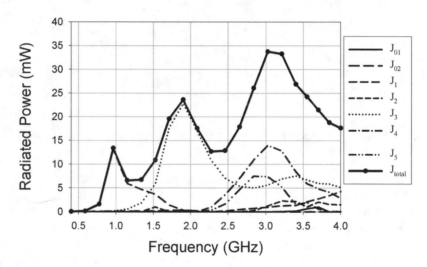

Figure 5.29 Contribution of the different modes to the total power radiated by the antenna.

of longitudinal modes J_3, J_4, and J_5. Note also that transverse mode J_2 is weakly coupled to the excitation.

Figure 5.30 shows the return loss obtained using the electromagnetic simulation software Zeland IE3D and measured for a fabricated prototype. As observed, both plots are very similar especially at the lowest frequencies, and the antenna is well matched at GSM and UMTS operating bands. Finally, Figure 5.31 illustrates the radiation patterns in the ZY and XY planes at 900 and 1,800 MHz. The omnidirectional behavior observed at both bands makes this antenna a good candidate for mobile handsets.

5.3.2.2 Coupling Element-Based Design of Mobile Terminals

Another alternative for the use of the radiating PCB modes consists of replacing the self-resonant antenna element of traditional mobile handsets by a nonreso-

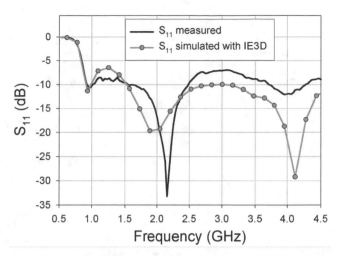

Figure 5.30 Simulated and measured return loss for the antenna prototype in Figure 5.29.

nant coupling element that efficiently excites the characteristic modes in the chassis terminal or PCB. A comprehensive study of the suitability of coupling elements for use in mobile terminals is provided in [62], where it is exposed that a considerable size reduction of the internal antenna can be achieved. Additionally, separate matching circuitry is needed to tune the combined coupling element-chassis structure in resonance. According to [62, 63], three parts can be distinguished in coupling element-based antenna structures, as shown in Figure 5.32. The chassis of the mobile terminal, the coupling element(s) and one or more matching circuitries.

In this scheme, the chassis is responsible for the radiation of the mobile terminal. Even in traditional mobile terminals with self-resonant antennas, the chassis has been found to be mainly responsible for the radiation at the lower frequency bands (at around 1 GHz) [13]. The coupling element is hence devoted to properly excite the characteristic modes in the chassis in order to increase its efficiency. The coupling element is not required to be resonant at the operating frequency, as it is not the main radiator anymore. Therefore, its size can be dramatically reduced, allowing for the design of very compact antennas for mobile terminals. Nevertheless, the location of the coupling element in the mobile terminal is very critical, since the excitation of the chassis modes is strongly determined by the coupling of the energy to the mode waveform, as discussed in the modal analysis presented in the previous section.

Efficiently coupling to the PCB characteristic modes requires locating the non-resonant element in a place where the electric field generated by the desired modes reaches its maximum, and hence where the current is at its minimum. In order to excite longitudinal current modes in the handset chassis, the optimum location of the nonresonant element lies near the short ends of the PCB, as derived from the current distribution of longitudinal modes shown in Figure 5.15. Different experiments on the influence of the coupling element location on the radiation have been performed in [13, 62] and they verify this statement.

Once the coupling element has been optimally placed above the PCB, a matching circuitry is required to match the impedance of the combined coupling ele-

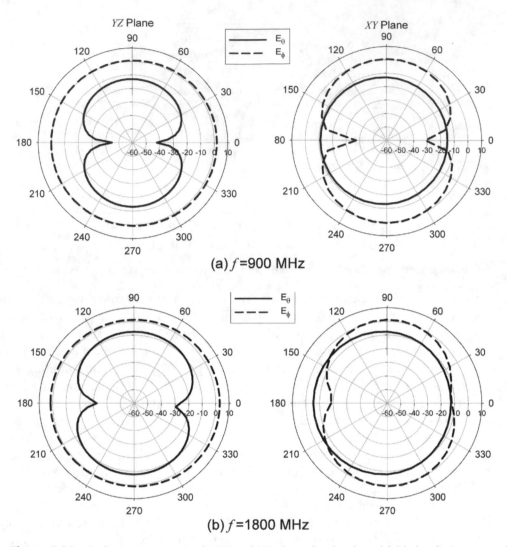

Figure 5.31   Radiation patterns in the YZ and XY planes for the slotted folded radiating ground plane (a) at 900 MHz, and (b) at 1,800 MHz.

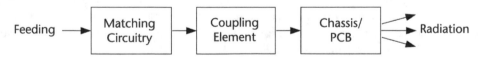

Figure 5.32   Scheme of a coupling element-based antenna structure.

ment-chassis structure to the traditional 50Ω transmission line, due to the fact that the coupling element is not working at its resonance frequency. With modern circuit technologies, one or more compact matching circuits can be easily designed and integrated in the mobile terminal without a substantial increment of the overall size. Multiband antennas can be obtained either by using multiple matching circuits having different matching frequencies, or by connecting a matching circuitry to different coupling elements designed for different frequencies [63]. Thus very compact multiband antennas for mobile terminals can be

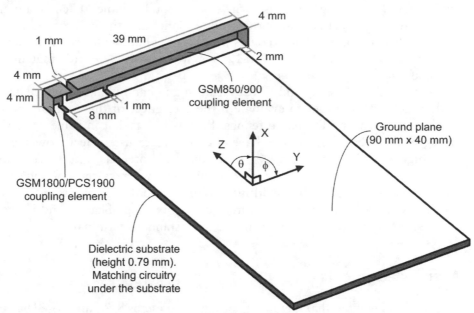

Figure 5.33 Geometries of the coupling elements and the PCB of an internal quad-band antenna [62]. Matching circuitry is not visible in the picture. (Reprinted with permission of John Wiley & Sons Inc.)

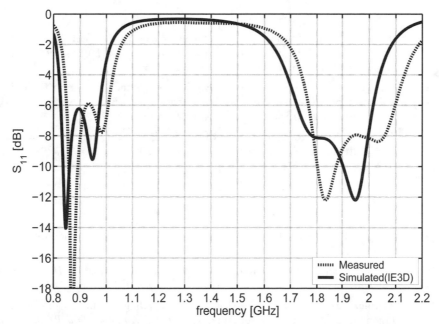

Figure 5.34 Simulated and measured return loss for the antenna presented in Figure 5.33 [62]. (From: [62]. Reprinted with permission of John Wiley & Sons Inc.)

obtained, although at the cost of a slight increase in the complexity of the associated circuits.

An internal quad-band antenna structure based on this concept is presented in [63]. The antenna geometry is shown in Figure 5.33. As observed, it comprises two resonant coupling elements with very low total volume ($0.7cm^3$), a multiresonant matching circuitry (under the ground plane) and the PCB of the mobile terminal. This compact antenna covers perfectly the bandwidth requirements of the GSM850/900 and GSM1800/PCS1900 systems with S_{11}<-6dB matching criterion, as shown in Figure 5.34.

Finally, the efficiency and SAR over head and hand phantoms have been studied in [62, 64], in order to evaluate if the user effect of this type of approach is higher than with traditional antennas. Results reveal that in some cases the coupling element-based structures have slightly higher user effect (lower efficiency and higher SAR) than the respective traditional antennas, like PIFAs. However, the difference cannot be considered significant, whereas the coupling elements have 3 to 6 times larger bandwidth-to-volume ratio than in PIFAs [65].

Consequently, this line of research represents a significant trend in the development of compact antennas for future multistandard terminals.

Acknowledgments

The authors want to thank Dr. Jaume Anguera (Fractus S.A.) and José Díaz Cervera for their valuable collaboration. This chapter benefits from the work carried out in the ACE Network of Excellence of the European Commission and the COST IC0603 Action "Antennas and Systems for Information Society Technologies."

References

[1]   Taga, T., and K. Tsunekawa, "Performance Analysis of a Built-In Planar Inverted F Antenna for 800 Mhz Band Portable Radio Units," *IEEE Journal on Selected Areas in Communications*, Vol. 5, June 1987, pp. 921–929.

[2]   Yamaguchi, R., et al., "Effect of Dimension of Conducting Box on Radiation Pattern of a Monopole Antenna for Portable Telephone," *IEEE Antennas and Propagation Society International Symposium Digest*, 1992, pp. 669–672.

[3]   McLean, J. S., "A Re-Examination of the Fundamental Limits on the Radiation Q of Electrically Small Antennas," *IEEE Transactions on Antennas and Propagation*, Vol. 44, May 1996, pp. 672–676.

[4]   Morishita, H., Y. Kim, and K. Fujimoto, "Design Concept of Antennas for Small Mobile Terminals and the Future Perspective," *IEEE Antennas and Propagation Magazine*, Vol. 44, No. 5, October 2002, pp. 30–43.

[5]   Wong, K. L., *Planar Antennas for Wireless Communications*, Wiley-Interscience, John Wiley & Sons, 2003.

[6]   Mittra, R., and S. Dey, "Challenges in PCS Antenna Design," *IEEE Antennas and Propagation Society International Symposium Digest*, 1999, pp. 544–547.

[7]   Martínez-Vázquez, M., M. Geissler, and D. Heberling, "Volume Considerations in the Design of Dual-Band Handset Antennas," *IEEE Antennas and Propagation Society International Symposium Digest*, July 2001, Vol. 2, pp. 112–115.

[8]   Ying, Z., "Some Important Innovations in the Terminal Industry in the Last Decade," *Proceedings of the 1st European Conference on Antennas and Propagation (EuCAP2006)*, ESA SP-626, October 2006, paper no. 244664.

[9]   Federal Communications Commission, Order on Recon and FNPRM, June 9, 2005, FCC 05-122, WT Docket No. 01-309, http://www.fcc.gov/cgb/dro.

[9abis] Federal Communications Commission, First Report and Order, February 26, 2008, FCC 08-68, WT Docket No. 07-250, http://www.fcc.gov/cgb/dro.

[10] Sato, K., et al., "Characteristics of a Planar Inverted-F Antenna on a Rectangular Conducting Body," *Electronics and Communications in Japan*, Vol. 72, No. 10, 1989, pp. 43–51.

[11] Manteuffel, D., et al., "Design Considerations for Integrated Mobile Phone Antennas," *Proceedings of the 11th International Conference on Antennas and Propagation*, Vol. 1, April 2001, pp. 252–256.

[12] Vainikainen, P., et al., "Performance Analysis of Small Antennas Mounted on Mobile Handsets," *Proceedings of the COST 259 Final Workshop–Mobile and Human Body Interaction*, 2000, p. 8.

[13] Vainikainen, P., et al., "Resonator-Based Analysis of the Combination of Mobile Handset Antenna and Chassis," *IEEE Transactions on Antennas and Propagation*, Vol. 50, No. 10, October 2002, pp. 1433–1444.

[14] Boyle, K.R., and L.P. Ligthart, "Radiating and Balanced Mode Analysis of PIFA Antennas," *IEEE Transactions on Antennas and Propagation*, Vol. 54, No. 1, January 2006, pp. 231–237.

[15] Manteuffel, D., A. Bahr, and I. Wolff, "Investigation on Integrated Antennas for GSM Mobile Phones," *Proceedings of the Millennium Conference on Antennas and Propagation AP-2000*, 2000, Vol. I, p. 304, paper no. 0587.

[16] Wu, T. Y., and K. L. Wong, "On the Impedance Bandwidth of a Planar Inverted-F Antenna for Mobile Handsets," *Microwave and Optical Technology Letters*, Vol. 32, No. 4, 2002, pp. 249–251.

[17] Kivekäs, O., J. et al., "Effect of the Chassis Length on the Bandwidth, SAR, and Efficiency of Internal Mobile Phone Antennas," *Microwave and Optical Technology Letters*, Vol. 36, No. 6, 2003, pp. 457–462.

[18] Arkko, A., "Effect of Ground Plane Size on the Free-Space Performance of a Mobile Handset PIFA Antenna," *Proceedings of the 11th International Conference on Antennas and Propagation*, Vol. 1, April 2003, pp. 316–319.

[19] Kivekäs, O., et al., "Bandwidth, SAR, and Efficiency of Internal Mobile Phone Antennas," *IEEE Transactions on Electromagnetic Compatibility*, Vol. 46, No. 1, February 2004, pp. 71–86.

[20] Hossa, R., A. Byndas, and M. E. Bialkowski, "Improvement of Compact Terminal Antenna Performance by Incorporating Open-End Slots in Ground Plane," *IEEE Microwave and Wireless Components Letters*, Vol. 14, No. 6, June 2004, pp. 283–285.

[21] Anguera, J., et al., "Enhancing the Performance of Handset Antennas by Means of Groundplane Design," *Proceedings of the IEEE International Workshop on Antenna Technology (IWAT 2006)*, March 2006, pp. 29–32.

[22] Anguera, J., et al., "Multiband Handset Antennas by Means of Groundplane Modification," *IEEE Antennas and Propagation Society International Symposium Digest*, June 2007, pp. 1253–1256.

[23] Lindberg, P., E. Öjefors, and A. Rydberg, "Wideband Slot Antenna for Low-Profile Hand-Held Terminal Applications," *Proceedings of the European Microwave Conference*, September 2006, pp. 1698–1701.

[24] Quintero Illera, R., and C. Puente Baliarda, "Multilevel and Space-Filling Ground-Planes for Miniature and Multiband Antennas," Patent Number WO 03/023900, March 2003.

[25]   Puente Baliarda, C., and J. Anguera Pros, "Slotted Ground-Plane Used as a Slot Antenna or Used for a PIFA Antenna," Patent Number WO2006097496, 2006.

[26]   Lindberg, P., and E. Öjefors, "A Bandwidth Enhancement Technique for Mobile Handsets Antennas Using Wavetraps," *IEEE Transactions on Antennas and Propagation*, Vol. 54, No. 8, August 2006, pp. 2226–2233.

[27]   Arkko, A. T., and E. A. Lehtola, "Simulated Impedance Bandwidths, Gains, Radiation Patterns and SAR Values of a Helical and a PIFA Antenna on Top of Different Ground Planes," *Proceedings of the 11th International Conference on Antennas and Propagation*, Vol. 2, April 2001, pp. 651–654.

[28]   Urban, R., and C. Peixeiro, "Battery Effects on a Small Handset Planar Inverted-F Antenna," *Proceedings of the IST Mobile & Wireless Communications Summit*, June 2004, Vol. 1, pp. 223–227.

[29]   Anguera, J., et al., "Multiband Handset Antenna Behaviour by Combining PIFA and Slot Radiators," *IEEE Antennas and Propagation Society International Symposium Digest*, June 2007, pp. 2841–2844.

[30]   Puente Baliarda, C., et al., *Multilevel Antennae*, Patent US 7015868 / US 7123208 / EP1526604, 2005-2006.

[31]   Anguera Pros, J., and C. Puente Baliarda, "Shaped Ground Plane for Radio Apparatus," Patent Number WO 2006070017, July 2006.

[32]   Peterson, A. F., S. L. Ray, and R. Mittra, *Computational Methods for Electromagnetics*, IEEE Press Series on Electromagnetic Wave Theory, New York: Wiley-IEEE Press, 1997.

[33]   Miller, E. K., L. Medgyesi-Mitschang, and E. H. Newman (eds.), *Computational Electromagnetics: Frequency-Domain Method of Moments*, IEEE Press, 1992.

[34]   Chong, E. K. P., and S. H. Zak, *An Introduction to Optimization*, Wiley-Interscience Series in Discrete Mathematics and Optimization, New York, John Wiley & Sons, 2001.

[35]   Rahmat-Samii, Y., and E. Michielssen, *Electromagnetic Optimization by Genetic Algorithms*, New York: John Wiley & Sons, 1999.

[36]   Christodoulou, C., M. Georgiopoulos, and C. Christopoulos, *Applications of Neural Networks in Electromagnetics*, Norwood, MA: Artech House, 2001.

[37]   Robinson, J., and Y. Rahmat-Samii, "Particle Swarm Optimization in Electromagnetics," *IEEE Transactions on Antennas and Propagation*, Vol. 52, No. 2, February 2004, pp. 397–407.

[38]   Pham, D. T., et al., "The Bees Algorithm: A Novel Tool for Complex Optimisation Problems," *Proceedings of the 2nd International Virtual Conference on Intelligent Production Machines and Systems (IPROMS 2006)*, 2006.

[39]   Garbacz, R. J., and R. H. Turpin, "A Generalized Expansion for Radiated and Scattered Fields," *IEEE Transactions on Antennas and Propagation*, Vol. 19, May 1971, pp. 348–358.

[40]   Harrington, R. F., and J. R. Mautz, "Theory of Characteristic Modes for Conducting Bodies," *IEEE Transactions on Antennas and Propagation*, Vol. 19, No. 5, September 1971, pp. 622–628.

[41]   Harrington, R. F. and Mautz, J. R., "Computation of Characteristic Modes for Conducting Bodies," *IEEE Transactions on Antennas and Propagation*, Vol. 19, No. 5, September 1971, pp. 629–639.

[42]   Garbacz, R. J., and D. M. Pozar, "Antenna Shape Synthesis Using Characteristic Modes," *IEEE Transactions on Antennas and Propagation*, Vol. 30, No. 3, May 1982, pp. 340–350.

[43]   Liu, D., R. J. Garbacz, and D.M. Pozar, "Antenna Synthesis and Optimization Using Generalized Characteristic Modes," *IEEE Transactions on Antennas and Propagation*, Vol. 38, No. 6, June 1990, pp. 862–868.

[44]   Harrington, R. F., and J. R., Mautz, "Control of Radar Scattering by Reactive Loading," *IEEE Transactions on Antennas and Propagation*, Vol. 20, No. 4, July 1972, pp. 446–454.

[45]   El-Hajj, A., K. Y. Kabalan, and R. F. Harrington, "Characteristic Modes Of A Slot in a Conducting Cylinder and Their Use for Penetration and Scattering, TE Case," *IEEE Transactions on Antennas and Propagation*, Vol. 40, No. 2, February 1992, pp. 156–161.

[46]   El-Hajj, A., and K. Y. Kabalan, "Characteristic Modes of a Rectangular Aperture in a Perfectly Conducting Plane," *IEEE Transactions on Antennas and Propagation*, Vol. 42, No. 10, October 1994, pp. 1447–1450.

[47]   Harrington, R. F., and J. R. Mautz, "Pattern Synthesis for Loaded N-Port Scatterers," *IEEE Transactions on Antennas and Propagation*, Vol. 22, No. 2, March 1974, pp. 184–190.

[48]   Mautz, J. R., and R. F. Harrington, "Radiation and Scattering from Bodies of Revolution," *Applied Scientific Research*, Vol. 20, June 1969, pp.405–435.

[49]   Austin, B. A., and K. P., Murray, "The Application of Characteristic-Mode Techniques to Vehicle-Mounted NVIS Antennas," *IEEE Transactions on Antennas and Propagation*, Vol. 40, No. 1, February 1988, pp. 7–21.

[50]   Harrington, R. F., *Field Computation by Moment Method*, New York: MacMillan, 1968.

[51]   Rao, S. M., D. R. Wilton, and A.W. Glisson, "Electromagnetic Scattering by Surfaces of Arbitrary Shape," *IEEE Transactions on Antennas and Propagation*, Vol. 30, No. 3, 1982, pp. 409–418.

[52]   Bai, Z., et al. (eds.), *Templates for the Solution of Algebraic Eigenvalue Problems: A Practical Guide*, PA Society for Industrial & Applied Math. (SIAM), 2000.

[53]   Cabedo, M., et al., "On the Use of Characteristic Modes to Describe Patch Antenna Performance," *IEEE Antennas and Propagation Society International Symposium Digest*, June 2003, Vol. 2, pp. 712–715.

[54]   Cabedo, M., et al., "A Discussion on the Characteristic Mode Theory Limitations and Its Improvements for the Effective Modeling of Antennas and Arrays," *IEEE Antennas and Propagation Society International Symposium Digest*, July 2004, Vol. 1, pp. 121–124.

[55]   Suter, E., and J. R. Mosig, "A Subdomain Multilevel Approach for the Efficient MoM Analysis of Large Planar Antennas," *Microwave and Optical Technology Letters*, Vol. 26, No. 4, August 2000, pp. 270–277.

[56]   Feick, R., et al., "PIFA Input Bandwidth Enhancement by Changing Feed Plate Silhouette," *Electronics Letters*, Vol. 40, No. 15, July 1004, pp. 921–922.

[57]   Antonino-Daviu, E., et al., "Wideband Antenna for Mobile Terminals Based on the Handset PCB Resonance," *Microwave and Optical Technology Letters*, Vol. 48, No. 7, July 2006, pp. 1408–1411.

[58]   Chang, F. S., and K. L. Wong, "A Broadband Probe-Fed Planar Patch Antenna with a Short Probe Pin and a Conducting Cylinder Transition," *Microwave and Optical Technology Letters*, Vol. 31, No. 4, November 2001, pp. 282–284.

[59]   Chen, Z .N., and M. Y. W. Chia, "A Feeding Scheme for Enhancing the Impedance Bandwidth of a Suspended Plate Antenna," *Microwave and Optical Technology Letters*, Vol. 38, No. 1, July 2003, pp. 21–25.

[60]   Jung, J. -H., H. Choo, and I. Park, "Small Broadband Disc-Loaded Monopole Antenna with Probe Feed and Folded Stripline," *Electronics Letters*, Vol. 41, No. 14, July 2005, pp. 788–789.

[61]   Kabacik, P., R. Hossa, and A. Byndas, "Broadening in Terminal Antennas by Tuning the Coupling Between the Elements and Its Ground," *IEEE Antennas and Propagation Society International Symposium Digest,* July 2005, Vol. 3A, pp. 557–560.

[62]   Villanen, J., et al., "Coupling Element Based Mobile Terminal Antenna Structures," *IEEE Transactions on Antennas and Propagation*, Vol. 54, No. 7, July 2006, pp. 2142–2153.

[63]   Villanen, J., C. Icheln, and P. Vainikainen, "A Coupling Element-Based Quad-Band Antenna Structure for Mobile Terminals," *Microwave and Optical Technology Letters*, Vol. 49, No. 6, June 2007, pp. 1277–1282.

[64]   Villanen, J., et al., "A Wideband Study of the Bandwidth, SAR and Radiation Efficiency of Mobile Terminal Antenna Structures," *Proceedings of the IEEE International Workshop on Antenna Technology (IWAT 2007)*, March 2007, pp. 49–52.

[65]   Vainikainen, P., et al., "Development Trends of Small Antennas for Mobile Terminals," *IEEE Antennas and Propagation Society International Symposium Digest,* 2007, pp. 2837–2840.

CHAPTER 6

Multiband Handset Antennas for MIMO Systems

Juan F. Valenzuela-Valdés, Antonio M. Martínez-González, and
David A. Sánchez-Hernández

6.1   Introduction

A strong research effort in the last decade has provided handsets with small, integrated multiband antennas with not-so-small radiation efficiencies. The possibility of a relatively large resonant patch antenna to reduce its size and attain diverse operating modes using slots [1] or spur lines [2] etched to the main patch has derived in hundreds of papers on multiband printed antennas and opened the way for planar inverted-F antennas (PIFAs). An alternative was to modify the main patch shape, obtaining C-shaped [3], U-shaped [4], T-shaped [5], E-shaped [6, 7], I-shaped [8], G-shaped [9], S-shaped [10], and many other alphabet-shaped antennas. Shorted and folded printed dipoles have also been reviewed in Chapter 3. Currently used multiband handset antennas typically contain a mixture of several basic multiband techniques. The conventional helices and monopoles, easily broken when in massive use, were changed and people noticed their handset antennas were "hidden." The technical specifications for handset multiband integrated antenna designs were getting a real challenge with a constantly reduced available volume. In the end, the original VSWR<2 requirement for evaluating operational bandwidth was relaxed for commercial handsets. This tremendous effort made possible, for instance, a nine-band integrated antenna [11], which in turn has made it possible to have all sorts of wireless systems (including GPS) available in our little handset. Yet, the technology push has also prompted new challenges for the handset antenna engineer. The presence of the user, antenna efficiencies, or simply optimum MIMO designs (including the MIMO cube) are all topics now being discussed. Achieving the high data peak rates specified in diverse wireless standards in a real system still remains as unlikely, despite the potential of handset MIMO to solve current interference and throughput problems of 3G and 3.5G. The new challenges, however, require diverse engineering disciplines to work closely together. Real-time, high-efficient signal-processing techniques, digital communications, microwave and radio-frequency (RF) engineering, channel propagation, and battery or antenna design become part of a unique and very complex design problem. It is somehow surprising that while many efforts have been and still are devoted to reduce MMIC loss by some fractions of a decibel, smart handset MIMO antennas have only recently gained attention. In contrast, smart handset MIMO antennas could really provide the extra tens of a decibel diversity gain needed to solve current coverage-spectral efficiency trade-offs of wireless systems. With the future advent of 4G, multiple-input multiple-output (MIMO) techniques are a hot research line, but the specific aspects of MIMO techniques when employed in a reduced volume such as the handset were yet to be fully addressed in a book. This chapter will identify the effects of the presence of the user, radiation, and mismatch efficiency on final MIMO capacity or diversity gain figures, and highlight the importance of antenna engineering in the upcoming new handset MIMO design processes. Likewise, novel MIMO techniques that are specifically ideal for handset design, such as the true polarization diversity (TPD) will be explained and compared to more conventional spatial, pattern, or polarization diversity techniques.

6.2   Design Considerations for MIMO Systems

The key feature of MIMO systems is their potential ability to turn multipath propagation from a pitfall into a benefit by multiplying transfer rates. The MIMO principle is simple and it is well understood with an analogy to the cable transmission. Should a transmitter send a signal through a cable, the capacity for transmitting information C is given by the well-known Shannon formula:

$$C(bits\,/\,s\,/\,Hz) = B \log_2\left(1 + \frac{S}{N}\right) \text{bits/s/Hz} \tag{6.1}$$

where S is signal power, N is noise power, and B is channel bandwidth. Should we have to send the same information through n different cables connecting n different transmitters to n different receivers, capacity would then become

$$C(bits\,/\,s\,/\,Hz) = n\,B \log_2\left(1 + \frac{1}{n}\frac{S}{N}\right) \text{bits/s/Hz} \tag{6.2}$$

This new capacity is inherently higher than the previous one, since the n factor placed multiplying outside the logarithm has a major impact than the same factor placed dividing within the logarithm. In this example the number of transmitters matches the number of receivers, but this does not always have to be the case. Since there are now n different propagation channels between each transmitter and receiver, an evaluation of the MIMO system through an H propagation matrix is required. The H matrix is known as the channel transmission matrix. The ijth element of H matrix represents the complex transfer function between the j transmitter and the i receiver. MIMO simply consists of adding the concept of multiplexing to propagation. Diversity gain concepts are employed when the different channels are used only to allow the designer the selection of the best propagation channel at each time slot. Should the different channels be combined somehow, the concept of MIMO capacity applies. Figure 6.1 shows an example of a MIMO system with two transmitting antennas and two receiving antennas (2×2).

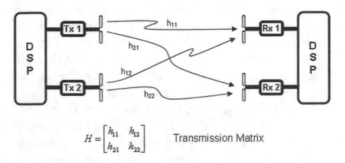

$$H = \begin{bmatrix} h_{11} & h_{12} \\ h_{21} & h_{22} \end{bmatrix}$$ Transmission Matrix

Figure 6.1   Example of a 2x2 MIMO system.

Despite the vast existing literature on MIMO techniques, the high ergodic simulated capacities early predicted for MIMO systems [12, 13] and recently confirmed for both correlated and uncorrelated scenarios [14–17], are blurred by the
diminishing effect of realistic channels [18]. The specific aspects of MIMO techniques when employed in a reduced volume such as the handset are yet to be fully
addressed in a book. While standard MIMO techniques do not have to consider
the presence of the user within the antennas' near field, handset MIMO inherently
contains this additional challenge. This poses numerous threats to the MIMO
designer, and novel techniques have appeared to overcome the effect of user presence on MIMO performance. The antenna engineer therefore faces novel upcoming challenges. With extraordinary efforts being currently placed on fabricating
efficient MMICs with low insertion loss, precise integrated phase shifters and
novel MEMs, handset fabrication may have to look back into the radiating structure for getting the 3 to 7 extra dBs that may be required.

Among the new techniques true polarization diversity (TPD) will receive particular attention. This technique has been reported to be able to nearly double
MIMO capacity for the same available volume. In this chapter, TPD will be
described and compared to more conventional spatial, pattern, or polarization
diversity techniques. In addition to TPD, another limiting factor for handset
MIMO systems in practice is the radiation efficiency. Large differences on the
measured correlation coefficients were observed in [19] due to the differences in
measured and ideal radiation efficiencies. For radiation efficiencies around 50%,
a variation of ±1 was expected in the correlation coefficients. These astonishing
figures may put handset MIMO in jeopardy. Basically, we know that integrated
antennas have an inherently low radiation efficiency [20], that the normalized
complex correlation coefficient (ρ_{rec}) always follows within the unit circle, that
uncorrelated branches are considered when their associated correlation coefficients are below 0.5, and that there are ±1 uncertainties when measuring the correlation coefficients due to radiation efficiencies. This means that handset
MIMO is not feasible. Or in other words, that it is not possible to get uncorrelated scenarios in the handset. It is obvious that more research is needed to this
respect. To begin with, the effect of radiation efficiencies has only been studied
through simulations [13] and with only two different branches under isotropic
environments. In this book chapter the works on the effect of measured efficiency on handset MIMO performance will be reviewed. Intimately linked to the
radiation efficiency is the problem of using MIMO techniques in the presence of
the user. With an ever-shrinking handset volume available for the radiating ele-

ments and the ever-increasing number of operating bands for the same element [21], the handset MIMO designer cannot avoid considering the user within the design. In this chapter several studies will be reviewed in relation to this.

Finally, it has been assumed for some time now that under noisy conditions that limit theoretical capacities there is no need to employ more receiving (R) than transmitting (T) antennas. In other words, the increase in MIMO capacity expected beyond $R>T$ is so small than it has been assumed that optimum MIMO performance is achieved for $T=R$. In this sense, increasing R beyond T will only have an effect on the diversity gain, but not on the capacity slope for a specific SNR [22]. This is one of the reasons why transmit diversity is commonplace while receive diversity use is scarce. However, this theory assumes that a large scattering area and a wide power angular spread exist around each terminal. In spite of the fact that the presence of the user increases local scattering, it is also true that it considerably reduces angular spread [23, 24]. Thus, handset MIMO may not have to follow some of the well-established conventional MIMO rules of thumb. In fact, recent studies [25, 26] have questioned the ergodic MIMO capacity limits since they do not account for some physical properties of the channel and the antennas. Recent studies regarding some new rules for handset MIMO will be described in this chapter.

In summary, the large profusion of studies and techniques about MIMO communications have increased our knowledge about this relatively new way of communicating. At the same time we are also exploring the communication performance limits. Therefore, the more we know, the more complicated the questions we ask ourselves become. In this chapter we will try to respond some of these questions.

6.2.1   Diversity Techniques

6.2.1.1   Spatial Diversity

Multipath fading is perhaps the major performance-limiting impairment on wireless communications. This problem arises mainly from destructive addition of multipaths in the propagation environment. Diversity techniques try to mitigate multipath fading by transmitting multiple replicas of the same information signal through as-independently-as-possible fading channel realizations. This scheme with multiple channel realizations is much less likely to fade than each individually, and higher spectral efficiency (total number of bits per second per hertz) is achieved. Some combination effort has to be performed at the receiver end. If only the best path is chosen, then the concept of spatial diversity applies. Otherwise, if a combination of all signals is employed in order to achieve a higher receive SNR, then MIMO capacity increase with respect to a single antenna defines performance. Selection (SC) and maximal ratio combining (MRC) are typically employed [27]. When the antennas are combined with full knowledge of the interference correlation between any two branches, adaptive beamforming or optimum combining can also be used. Because MIMO systems use antenna arrays, interference can be mitigated naturally with the spatial distance that separates each antenna in the array. The first technique employed to increase capacity was then to use an array of spatially separated antennas, thus the term spatial diversity was coined. The earliest form of spatial transmit diversity in wireless communica-

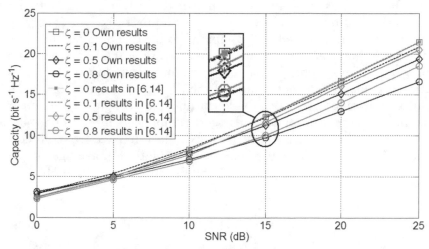

Figure 6.2 Simulated [14] and measured 3x3 MIMO capacity with spatial diversity techniques versus SNR (dB).

tions was the delay diversity scheme proposed in 1993 [28], where a signal was transmitted from the second antenna, then delayed one time slot, and transmitted from the first antenna. In 1993 this technique was known to be employed for quite some time on the early unidirectional hyperbolic radionavigation systems such as GEE, LORAN, and DECCA, among others, where the master-slave scheme was based upon slaves being triggered by the incoming master pulse. Since there is a certain separating distance in spatial diversity, mutual coupling exists between elements, and some correlation is expected to degrade the estimated increment in capacity. In early MIMO systems, spatial correlation figures were sufficient data to estimate MIMO capacity for simple fading models.

Figure 6.2 shows the simulated [15] and measured ergodic 3×3 MIMO capacity under Rayleigh-fading environments for diverse correlation coefficients. Figure 6.3 illustrates the relationship between the correlation coefficient and the wavelength-normalized dipole spatial separation D (d/λ) versus spatial diversity gain obtained through measurements in a reverberation chamber.

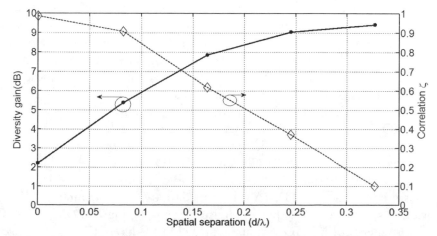

Figure 6.3 Measured correlation and spatial diversity gain for a 3x2 MIMO system.

These figures show that in spatial diversity schemes the additional diversity gain provided by a correlation coefficient ζ lower than ~0.5 (corresponding to a spatial separation of D=0.24) compared to the one obtained for ζ=0.5 is not significant [14, 15, 22, 29]. Figure 6.4 depicts a typical GSM base station antenna in one sector employing spatial diversity. In the multipath fading scenario, the Ricean K factor defines the ratio of the fixed and variable components power. In this way it reflects the contribution of the line of sight (LoS) components in LoS channels or the deterministic strongest components in non-LoS (NLoS) scenarios to the total channel gain. The K factor can be estimated from all received data for each MIMO channel using the moment method [30], averaged over its all corresponding single-input single-output (SISO) subchannels by [31]

$$K = \sqrt{1-\gamma} \, /(1 - \sqrt{1-\gamma})$$
(6.3)

where $\gamma = \sigma_r^2/P_r^2$, and σ_r^2 is the variance of the received signal power about its mean P_r. When no direct coupling (perfect rich multipath) exists between the transmitter and the receiver, the received signal follows a Gaussian distribution, and therefore its envelope a Rayleigh distribution, with an uniform phase distribution over 2π. Spatial diversity defined above has already been proven as an effective way to overcome multipath fading. However, increasing the number of antennas with spatial diversity will not linearly increase the capacity for users under high K-factor conditions (non-Rayleigh). Since the correlation between the signal replicas depends on both the antenna spacing and the angle spread of the wireless propagation channel, and since on the base station side the angle spread is typically narrow (10° to 20°), at least 20λ horizontal and 15λ vertical separation distances are required for efficient spatial outdoor diversity in practice. In the handset, the main constraint for applying spatial diversity is that the required space is not available. At the mobile station the angle spread is wide, typically close to 360°, and decorrelation can be achieved with spacings of the order of a quarter wavelength. With the constantly shrinking handset volume, this is still a hard challenge for the antenna designer. In addition, the mobile station needs to remain low-cost and its battery life must not be compromised. As far as we know only the Japanese Personal Digital Cellular (PDC) system employs spatial diversity at the mobile handset. PDC employs dual-antenna diversity on handsets. The diversity scheme employs a simple switching network to select the antenna that yields the highest SNR out of an array of two antennas. Japanese operator NTT DoCoMo also offers handsets with dual antennas on its freedom of mobile multimedia access (FOMA) 3G network. Beyond 3G, NTT DoCoMo has included MIMO in its 4G standard since a successful test of multiple-antenna systems transmitting at spectral efficiencies of 25 bits/s/Hz were announced in 2006. The company reached nearly 5 Gbps in a MIMO test for its 3.9G Super-FOMA using single-carrier frequency division multiple access (SC-FDMA) in the uplink. Despite its name, SC-FDMA transmits data over the air interface in many subcarriers, as orthogonal frequency division multiplexing (OFDM) does. Commercial deployments of MIMO systems worldwide are likely to take place sooner rather

Figure 6.4   Spatial diversity in a typical urban GSM base station.

than later. Recent advances in battery technology, which included the use of micro fuel cells in handsets, have solved the video requirements of 3.5G systems, and highly efficient space-time processing (STP) algorithm are now commonplace. Thus, the possible advantages of handset MIMO will have a place in 4G systems. Upcoming 4G wideband orthogonal frequency division multiplexing (W-OFDM), for instance, will undoubtedly include smart handset MIMO antennas. Yet, the use of MIMO in close proximity of the user and within a reduced volume poses numerous questions on the suitability of this technique. With the obvious drawbacks of spatial diversity in handsets, hybridization of diverse techniques leads current research efforts.

6.2.1.2 Pattern Diversity

Spatial diversity requires space, which is not always available. To introduce significant subchannel decorrelation to the MIMO system that has insufficient space available the MIMO designer can make use of pattern diversity [32]. Pattern diversity, also known as angle diversity, provides the required diversity gain by using different beams. Significant improvements are shown in system performance in both Rayleigh and Rician fading environments when antennas with different radiation patterns are employed [33]. Diversity gain can be explained by the fact that different antenna patterns receive different sets of multipath waves. This gain will be largest if the two sets are completely independent that can, in theory, be realized with two antenna patterns that have zero overlap [34]. Figure 6.5 reproduces the comparison between spatial-only and a combination of spatial and pattern diversity in [34]. For two different patterns combined to two 0.1λ spatially separated antennas (to reproduce the situation in a handset), an extra gain of 1 dB was accomplished. Should 25 different patterns be used, predicted extra gains ranged from 3 to 6 dB. The benefit, however, seems to increase as the multipath environment becomes more complicated. Pattern diversity also behaves

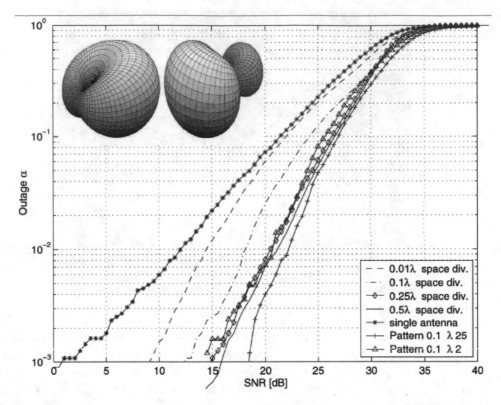

Figure 6.5   Measured CDFs for dual selection combining for different antenna spacings and antenna-pattern diversity. (From: [34]. ©2004 IEEE. Reprinted with permission.)

like spatial diversity in the sense that the largest performance improvement occurs when this first diversity branch is added [33]. The performance improvements are achieved at the cost of antenna complexity.

Some authors define pattern diversity as the natural generalization of conventional orthogonal polarization diversity since the orthogonality between patterns decorrelates the signals in highly scattering environments [35]. Some authors have identified pattern diversity to have better performance than spatial diversity when used in handsets [34]. However, this may be limited to scenarios with relatively large angular spread, such as urban and dense urban environments, making its use on rural scenarios somehow limited.

6.2.1.3 Polarization-Diversity

Despite the advantages of pattern diversity, with the spatial requirements for efficient spatial diversity the possibility of getting extra channels by using orthogonal polarization states has recently gained attention. This is known as polarization diversity. The improvement granted by polarization diversity in wireless systems is typically obtained by an additional decorrelated channel provided by a polarization state made orthogonal to the existing one, usually at the transmitting end. A randomly oriented linearly polarized antenna is also typically used at the receiver for evaluating polarization diversity. In this scheme the cross-polarization discrimination (XPD) factor is the usual evaluating parameter, with low correla-

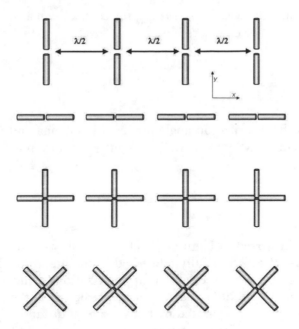

Figure 6.6   Polarization diversity.

tion coefficients being achieved even in NLoS situations [36]. While keeping the orthogonal characteristic, several options are currently being used, such as the crossed dipoles or 45° slanted configurations depicted in Figure 6.6. Some combinations of two-branch orthogonal polarization and spatial diversity have been reported [37]. Recently, a triaxial combination of polarization and pattern diversity has also been proposed [38–42].

6.2.1.4 True Polarization Diversity

In mobile communications scenarios, however, multiple scattering may not be sufficient for a given polarization to decouple half its power into the orthogonal polarization [14]. Moreover, both reflection and diffraction processes are polarization sensitive. Measurements have already demonstrated that copolar and cross-polar components in typical outdoor environments are almost uncorrelated [43]. Channel behavior is therefore different for different polarization states [40] and fading cross-correlations may even increase the ergodic capacity beyond the case of independent channels [44]. Although very effective, conventional orthogonal polarization diversity alone does not suffice once the number of antennas exceeds the number of orthogonal polarizations, which is very small [45]. This enhances the potential of using multiple polarization states to avoid the possible lack of richness in multipath. True polarization diversity (TPD) is defined [46] by rotating a certain angle one antenna with regard to a contiguous one in the MIMO array. In this way an arbitrary angular separation between contiguous dipoles is employed in an equivalent way that an arbitrary spatial separation is employed for spatial diversity. The polarization state can be defined by the inclination angle respect to the vertical dipole ϕ_n. Should the different polarization states be limited between 0 and α_{max}, for each antenna, the polarization angle of each antenna can be calculated by,

$$\phi_n = \frac{(n-1) * \alpha_{max}}{N} \qquad (6.4)$$

where ϕ_n is the polarization angle for the nth antenna and N is the total number of antennas in the MIMO array. The angular separation is defined by

$$d\theta = \phi_n - \phi_{n-1} \qquad (6.5)$$

TPD is illustrated in Figure 6.7. The possibility of employing many different polarization states is not fully addressed in the literature. This is mainly due to (de)coupling effects between different polarizations being a complex mechanism to be simulated [47]. In fact, an accurate prediction of the correlation coefficient between two dipoles separated by both a spatial distance and an arbitrary angular position has not been available until very recently [48]. For only two receiving antennas, it seems obvious that the best option is to employ two orthogonal polarization states. Figure 6.8 depicts the measured ergodic capacity of i.i.d Rayleigh-fading 3×2 MIMO channels versus *SNR* with an angular separation $d\theta$ between contiguous dipoles as a parameter and *D*=0. This situation (*D*=0, *R*=2) is not realistic unless a multimode antenna is employed, yet it is extremely important for comparison purposes. From Figure 6.8 it is easily observed that for angular separation values over 54°, the additional increase in terms of MIMO capacity is negligible. It is indeed true that the orthogonal configuration gives the best results for a two-branch scheme, but a difference of only 0.4 bits/s/Hz in MIMO capacity was encountered between an angular separation of 54° and that of 90° for a SNR=10 dB. Figure 6.9 illustrates the effect of angular separation ($d\theta$) and correlation coefficient (ζ) on measured diversity gain for 3×2 TPD-only MIMO schemes.

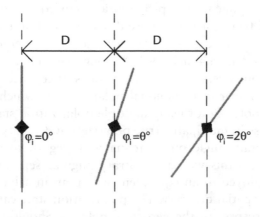

Figure 6.7   True polarization diversity in a MIMO array. (From: [46]. ©2006 IEEE. Reprinted with permission.)

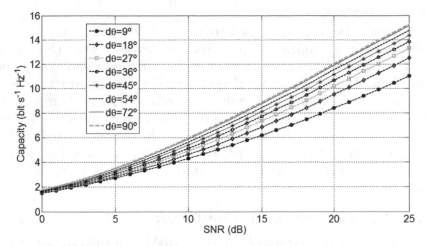

Figure 6.8   Measured 3×2 MIMO capacity versus SNR with $D=0$.

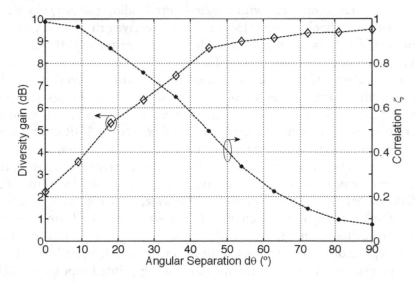

Figure 6.9   Measured 3×2 MIMO correlation and TPD gain versus angular separation ($d\theta$).

At least for Rayleigh-fading environments, true polarization diversity performs in a similar way to spatial diversity, even with just a two-branch scheme. As an example, a spatial separation of ~0.14λ is equivalent to an angular separation of ~36° for a 3×2 MIMO system. In fact, true polarization diversity also shows a limit in angular separation values above which diversity gain increase is negligible. The limit was found to be $d\theta=54°$. For $d\theta=54°$ the measured correlation coefficient ζ is ~0.5, which is consistent with the results obtained for spatial diversity. Yet, unlike what happened for spatial diversity, the polarization diversity inherent limitation (a maximum of 360° are available for the whole array) has an effect on diversity gain for larger MIMO arrays. In spite of this limitation, when more than two branches are employed and for reduced volumes, TPD outperforms conventional orthogonal polarization diversity [49].

6.2.1.5   Frequency Diversity

Since the MIMO channel has a frequency-dependent behavior, a choice of frequency-spaced channels also provides some diversity gain. This type of diversity is typically employed as frequency hopping for point-to-point or mobile communications. The condition for channel spacing, also known as coherence bandwidth, is to obtain sufficiently uncorrelated fading branches. Should the frequency spacing be smaller than the coherence bandwidth, then the diversity gain would be reduced due to correlation. The coherence bandwidth is inversely proportional to the delay spread. The advantage of this technique is that the use of multiple antennas is avoided, provided that a single antenna attains sufficient operational bandwidth.

6.2.2   Volume Constraints: The MIMO Cube

Volume constraints suggested a compact MIMO antenna consisting of 12 dipoles placed in the edges of a cube [50]. In this scheme both spatial and polarization diversity were combined, with the box surrounding the antenna not larger than that for three dipoles. The MIMO cube can receive or transmit waves of arbitrary polarizations from all possible 3-D directions. For the MIMO cubes, there are 12 electric dipole antennas at the 12 different edges of the transmitter and the receiver cubes, and therefore, the transmission matrix H consists of 12×12 transmission links. The dipoles are oriented in all three dimensions x, y, and z, and all axes, making use of the tripolarization technique. Up to 8.5 dB highest mean gain per cube and a theoretical capacity of 62.5 b/s/Hz for SNR=20 dB were obtained in the simulations. This capacity value is close to a theoretical i.i.d. 12×12 Rayleigh case. The slight difference is due to the existence of cross-coupling (CxP) between elements. Even when the available volume is reduced to a side length of $\lambda/20$, the MIMO cube still provides 34 b/s/Hz, which is more than five times that of a single Rayleigh fading channel (5.9 b/s/Hz). The performance of the MIMO cube for different side lengths is reproduced in Figure 6.10. Mutual coupling has been reported to have an important effect for side lengths below 0.3λ [51]. The first measurements are already available using printed dipoles in [52], also illustrated in Figure 6.10. Some pattern diversity was also found since different radiation pattern were observed [52]. Using just a $(0.5\lambda)^3$ volume, the achieved MIMO capacity was 55.9 b/s/Hz for a SNR=20 dB, only 9.9 b/s/Hz less than initially predicted. Some impedance shifting was observed. As has been well explained by several papers [26, 52], the practical implementation of the MIMO cube will still have to face some additional problems, such as mutual coupling, feeding schemes or efficiency, among others, that reduce the predicted capacity figures and shifts the operating frequency range.

6.3   MIMO Techniques in a Handset

While MIMO techniques have received considerable attention in the literature with important throughput increments, handset MIMO is not yet a reality. The reasons behind this somehow contradictory reality lay on the inherent difficulties

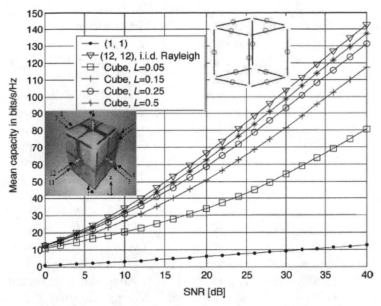

Figure 6.10   The MIMO cube concept. (After: [26, 52]. ©2006 IEEE. Reprinted with permission.)

for handset MIMO in practice and the important technology changes that handset MIMO requires. Being technology-driven, handset MIMO will consequently increase the complexity and cost of the mobile station. In addition, some handset MIMO design challenges are still to be fully addressed in the literature. In real operating handset MIMO the presence of the user and the antenna efficiency can no longer be ignored, and its effects are important not only for antenna and handset design, but also for MIMO performance. Therefore, some of the issues inherently linked to multiband handset MIMO in practice will be reviewed.

6.3.1   Effect of Antenna Efficiency

The diverse factors affecting diversity gain [23, 53–56] and MIMO capacity [14, 56] have been deeply but independently studied in the literature, with the general aim of developing specific techniques for a reduction of the correlation coefficients between signals. In most studies, however, ideal efficiencies are employed for the transmitting and receiving antennas. However, recent research results have highlighted the role of efficiency for calculating the correlation coefficients between signals received on any two antennas using single-measurements S-parameters in an isotropic signal environment [19]. A large degree of uncertainty, highly related to the difference between the ideal efficiency and the real efficiency, is attributed to the role of efficiency on final diversity performance. With 50% antenna efficiencies achievable, uncertainties in the normalized complex correlation coefficient due to the radiation patterns (ρ_{rec}) are up to ±1. With the typically low antenna efficiencies found in the handset scenario [20], since ρ_{rec} is always within the unit circle and since two diversity branches are considered to be uncorrelated when the correlation coefficient is less than 0.5 [57], current state-of-science would lead to the impossibility of obtaining uncorrelated channels. This would in turn mean that neither high diversity gain nor high MIMO capacity could be achieved at the handset scenario with two low-efficient branches. Yet, the importance of radiation efficiency in [19] also outlined large theoretical inaccuracies and was only studied through simulations for a two-branch diversity scheme within isotropic signal environments. Handset MIMO could not be a no-way street.

In addition, it has been assumed for some time now that under noise-limiting conditions and despite theoretical MIMO erdogic capacity formulas, there is no need to make $\beta = \frac{T}{R} > 1$, where T and R are the number of transmitting and receiving antennas, respectively [56]. Namely, there is no need to use more receiving antennas than transmitting antennas for MIMO systems, and when keyhole or some type of rank deficiency is present, increasing the number of transmitting or receiving antennas may only have an effect on diversity gain but not on the slope of MIMO capacity curves over a signal-to-noise ratio (SNR) [22]. Yet the MIMO designer cannot expect capacity to remain unaffected by correlation coefficients or efficiency for more general $m \times n$ MIMO systems, wherein accurate predictions of the correlation coefficients for real propagating environments are still required for deriving precise system capacity. This is particularly important at the receiver end (receive diversity) due to both the inherent proximity of the user and the commonly low values of antenna efficiencies when antennas are integrated in a limited volume [20]. Moreover, correlation coefficients alone do not tell the whole story for MIMO performance since channel capacity depends on each instantaneous realization of the channel matrix, angular spread available and not just average correlation properties [21, 24]. A good effort that kept the handset MIMO door open was described in [55]. Efficient (>50%) and complementary antennas with very low correlation coefficients (<0.12) were designed and fabricated in [55], and the authors concluded that receive diversity with a relatively large number of antennas located in a volume-limited space was still an open issue since efficiency and mutual coupling play an important role. The key for handset MIMO success lays on these two variables.

Radiated power in handset MIMO is the power available for communication. It can be determined by integrating the normal component of the time-averaged Poynting vector over a closed surface, which encloses the antenna and the user

$$P_{rad} = \oint_{\partial V} P_{\partial V} dS = \oint_{\partial V} \frac{1}{2} * \text{Re}\{E \times H*\} dS \tag{6.6}$$

where E and H are the electric and magnetic field vectors (amplitudes). The radiated power can also be defined as

$$P_{rad} = P_{max} - P_{tloss} = P_{av} - P_{diss} - P_{body} - P_{emb} - P_{m} \tag{6.7}$$

where

- P_{max} Maximum power available at the antenna input
- P_{tloss} Total power loss

- P_{diss} Ohmic and dielectric power loss
- P_{body} Power absorbed in the human body
- P_{emb} Power dissipated in other antennas through mutual coupling
- P_m Power loss due to impedance mismatch

The power absorbed in the user P_{body} can be found either by the surface integral of the normal component of the time-averaged Poynting vector entering the surface of the lossy scatterer or by integrating the specific absorption rate (SAR) over the entire volume of the scatterer (user)

$$P_{body} = \oint_{S_{SCAT}} P_{\partial V} \, dS_{scat} = \oint_{V_{scat}} \rho \cdot SAR \, dV_{scat} \qquad (6.8)$$

where ρ is the density of the tissue and SAR is the specific absorption rate in W/Kg. Significant gain reduction occurs when an antenna is used close to the human body, with resulting radiation efficiencies below 50% [58]. When the antenna is integrated in a small volume, an additional efficiency reduction is also expected [20], aggravating potential MIMO performance. Consequently, a decrease in the mean effective gain (MEG) is expected, yet diversity gain remains nearly the same [24], making the study of effective radiation efficiency for diverse efficiencies, effective mean effective gain (EMEG) [24], and MIMO capacity essential for properly studying the effect of efficiency on receive diversity and MIMO performance.

The total embedded radiation efficiency η_{ef} is the ratio between the radiated power P_{rad} the total available power

$$\eta_{ef} = \frac{P_{rad}}{P_{rad} + P_{body} + P_{diss} + P_{emb} + P_m} \qquad (6.9)$$

A good example of employing the total embedded radiation efficiency for overcoming the effects of the innate low efficiency of handheld antennas can be found in [59]. Some authors have included mutual coupling in the calculation of the total embedded radiation efficiency [58] while others tend to separate these two effects since mutual coupling in wireless communications is not constant and depends strongly on multipath scenarios and scattering objects [60]. In order to identify only efficiency-related effects, mutual coupling has to be excluded when calculating the total embedded radiation efficiency. This can be done by making P_{emb} equal to zero, in a similar way to recent switched array techniques [60], wherein the advantage of eliminating mutual coupling is achieved by using a single transmitter and a single receiver to measure the transfer function in a sequential arrangement at the cost of long channel mean stationary times. In this way the effect of efficiency is isolated from others that have an effect on the matrix correlation properties.

When taking efficiency into account, a new definition for diversity gain was developed in [49]. For the apparent diversity gain (ADG) the reference is the strongest branch/antenna while for the effective diversity gain (EDG) the reference is the calibration antenna to avoid discrepancies in measured results due to the spacing-dependent reduction in radiation efficiency that occurs when spacing-dependent mutual coupling exists [61]. The ideal diversity gain (IDG) was defined when the reference is the theoretical upperbound Rayleigh curve [49]. The effective diversity gain depends on the radiation efficiency of the reference antenna while the ideal diversity gain does not depend on this radiation efficiency. This new term helps identifying efficiency-related effects, not only for diversity gain, but also for MIMO systems. In fact, while diversity gain for multiple antennas has been studied from different points of view depending on many factors [23, 53–55] (number of antennas, correlation, multipath scenario, algorithm used to combine the different signs, and radiation efficiency, among others), existing studies on MIMO capacity including efficiency effects are scarce [49]. Furthermore, contradictory findings can be found for the effect of radiation efficiency on diversity gain. In [58] a reduction in efficiency when the user was present did translate in a large increase in the mean effective gain (MEG) for both a $\lambda/4$-whip antenna and a $\lambda/2$-PIFA antenna, but a relatively small increment for a $\lambda/4$-PIFA element, mainly attributed to a change in the direction of the incident waves. In contrast, in [21] MEG was halved when the upper body was present for both the vertically and horizontally polarized monopole and PIFA antennas. In fact, it is now clear that MEG is strongly dependent upon both antenna [62] and user [63] characteristics, and that other parameters such as the diversity antenna gain (DAG), which includes the correlation coefficient and the MEG characteristics, or MIMO capacity, are required for full performance assessment of MIMO systems. The effective diversity gain is calculated as,

$$EDG(dB) = IDG(dB) + \eta_{ref}(dB) \tag{6.10}$$

where η_{ref} is the efficiency of the reference antenna. EDG is inherently less accurate than IDG since it depends upon the efficiency of the antenna reference, which usually has an uncertainty component. With the introduction of IDG another term can immediately be coined for the employed combining scheme of the N receiving antennas; the diversity gain loss (DGL) as

$$DGL_N(dB) = IDG(dB) - sel\_comb_N(dB) \tag{6.11}$$

If the signals are uncorrelated, this new term also can be defined as

$$DGL_N(dB) = \frac{-\sum_{i=1}^{i=N} \eta_i(dB)}{N} \tag{6.12}$$

Figure 6.11 and Table 6.1 illustrate the concepts of ADG, EDG, IDG, and DGL for a two-branch diversity scheme using η_i=0.91 dipoles on a reverberation chamber. From this figure we can observe that the effect of efficiency on the CDF

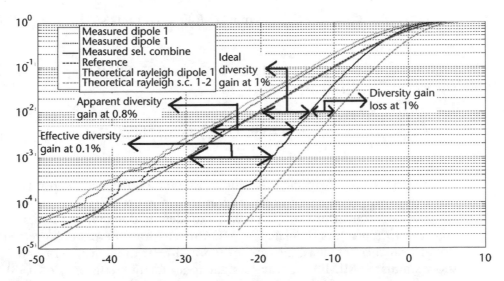

Figure 6.11 Measured diversity gain cumulative probability density functions (CDF) versus relative received power level (dB) for two-parallel dipoles with 0.5λ spacing and η_i=0.91.

Table 6.1   A Comparison of Different Diversity Gain Definitions

	EDG (dB)	IDG (dB)	ADG (dB)
At 1%	7.3	7.0	9.5
At 0.5%	8.5	8.0	11.0
At 0.3%	9.8	9.2	12.6
At 0.1%	12.1	12.5	16.1

of diversity gain is a displacement of the curve to the left, but the slope is not affected. In order to evaluate the effect of the total embedded radiation efficiency on diversity gain, we can write

$$SNR_{we}(dB) = SNR(dB) + \eta_{ef}(dB) \tag{6.13}$$

where SNR_{we} is the signal to noise ratio with the efficiency effect included. Figure 6.12 depicts the simulated cumulative probability density functions for different total embedded radiation efficiency with different numbers of uncorrelated branches/antennas with the same total embedded radiation efficiency. It can be observed from this figure that numerical results agree well with (6.12) and (6.13), with unaltered capacity slopes when efficiency is accounted for. The validity of (6.12) is also confirmed by Figure 6.12, with DGL values obtained as the average of the different antenna efficiencies.

Regarding MIMO capacity, the radiation efficiency was accounted for in the corresponding equation in [49] by

$$C_{MIMOwe} = C_{MIMO} + \min\left(T,R\right)\log_2\left(\eta\right) \text{ bits/s/Hz} \qquad (6.14)$$

In a similar way to the DGL, we can obtain the MIMO capacity loss due to efficiency (CL_{MIMOef}) as

$$CL_{MIMOef} = -\min\left(T,R\right)\log_2\left(\eta\right) \text{ bits/s/Hz} \qquad (6.15)$$

From (6.15), the capacity loss due to radiation efficiencies depends upon the number of transmitting and receiving antennas and their radiation efficiencies, and it is not straightforward. In an effort to identify dominating factors in (6.15), Figure 6.13 represents simulated MIMO capacity for different uncorrelated number of branches/antennas, all with η_i=0.5 and β=1. In Figures 6.14 and 6.15 we show simulated MIMO capacity loss due to a radiation efficiency of η_i= 0.5 with different numbers of branches/antennas versus SNR. Ninety pecent of the capacity loss due to radiation efficiency is reached in all studied systems for a SNR=15 dB. Lower efficiencies require higher SNRs for achieving 90% of the capacity loss due to radiation efficiency. From these simulations it seems clear that efficiency plays an important role for low SNRs since capacity loss is practically constant above SNR=35 dB for all combinations. Figure 6.16 depicts simulated MIMO

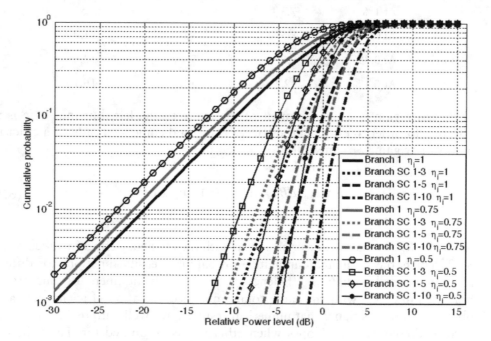

Figure 6.12   Cumulative probability density functions for different numbers of uncorrelated branches with diverse total embedded radiation efficiency.

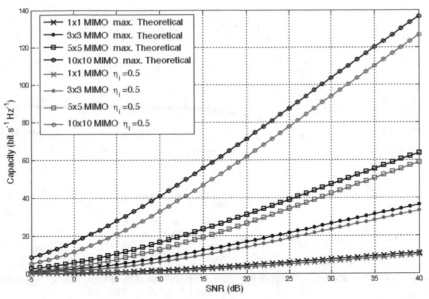

Figure 6.13 MIMO capacity for a different number of branches versus efficiency.

capacity loss due to radiation efficiency for different systems with two different SNR (5 and 20 dB), $T=3$ and R as parameter. MIMO capacity loss due to efficiency increases linearly with increasing number of receiving antennas until $T=R$, from which the rate is reduced.

Capacity loss reaches 90% of its maximum when $R=T$ for all studied systems, making additional capacity loss due to efficiency for $R>T$ MIMO systems negligible compared to $R=T$ MIMO systems. Yet MIMO capacity when accounting for antenna radiation efficiencies depends strongly on SNR, the number of antennas and their efficiency. Consequently, adding a high efficient antenna may enhance

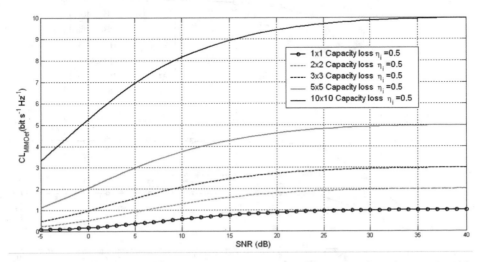

Figure 6.14 $m \times m$ MIMO capacity loss versus SNR for different number of antennas with $\eta_i=0.5$.

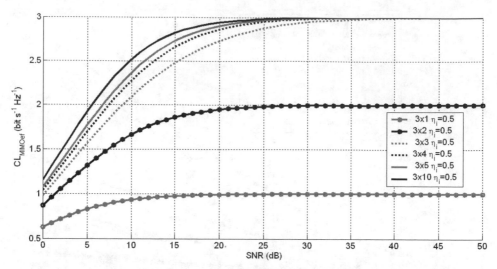

Figure 6.15   3 x *n* MIMO capacity loss versus SNR for different number of antennas with η$_f$=0.5.

final MIMO capacity when radiation efficiencies are accounted for, which always happens in practice.

Measured 3×10 MIMO selective combining diversity gain can be observed in Figure 6.17. Each branch is displaced to the left in a value proportional to its radiation efficiency. In Figure 6.18 we show the cumulative probability distribution function for four 10-branch combining systems with different radiation efficiencies. Figure 6.19 illustrates the evolution of EDG for the simple combining situations of up to 10 receiving antennas with the same radiation efficiencies. Increasing the number of receiving antennas has a positive effect on EDG, but

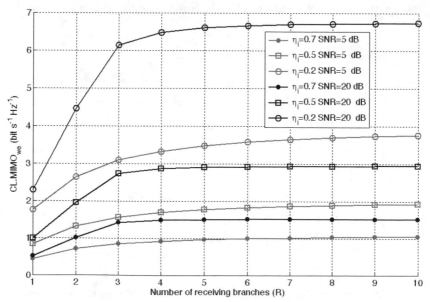

Figure 6.16   Simulated MIMO capacity loss for different radiation efficiencies versus *R* with *T*=3. (From: [49]. ©2008 IEEE. Reprinted with permission)

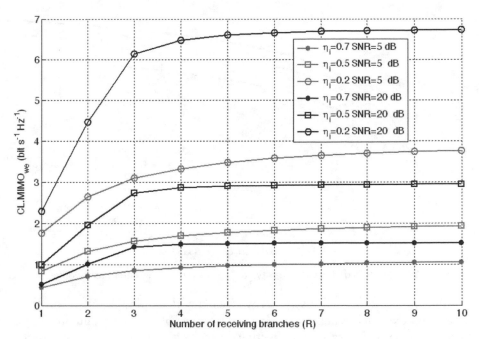

Figure 6.17 Cumulative probability distribution function of measured values in the reverberation chamber for different radiation efficiencies at 2,400 MHz.

from $R=T$ onwards the increment rate (slope) slows down considerably. For a $T=3$ system with 0.36 antenna efficiencies, increasing the number of receiving antennas from 2 to 3 attains an additional 5-dB EDG, which is exactly the same additional effective diversity gain that can be obtained when adding seven extra antennas to the $R=3$ situation. Better results, however, are observed when adding more efficient antennas. For instance, the same $T=3$ system but with 0.81 antenna efficiencies attains 4 dB additional EDG when adding an additional antenna to an $R=2$ system, but approximately 6 dB are gained when seven antennas are added

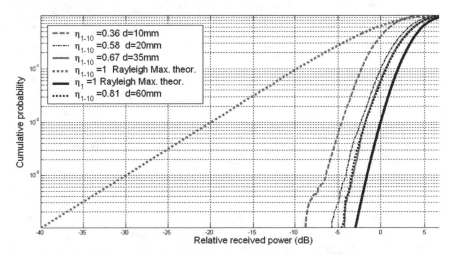

Figure 6.18 Cumulative probability distribution function for different measured combining systems.

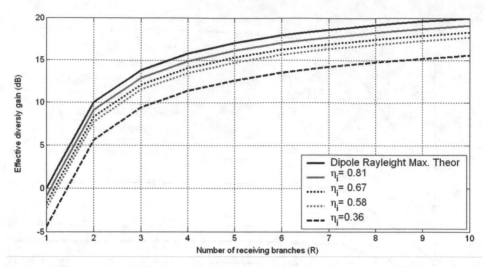

Figure 6.19   Measured effective diversity gain versus R receiving antennas for different efficiencies.

to the $R=3$ situation. Consequently, efficiency does play an effect on the increasing EDG expected when increasing the number of receiving antennas, apart from simply providing worse final effective diversity gain values compared to ideal lossless antennas, even when all receiving antennas exhibit the same radiation efficiency. Figure 6.20 shows different measured combining scenarios [49].

Figures 6.21 and 6.22 exhibit MIMO capacity and capacity loss due to efficiency, respectively, for four different MIMO systems with the same radiation efficiency for all receiving antennas, but different effective radiation efficiency for each system. These simulations confirmed that the slope of the capacity loss due to efficiency depends on T and R. It is also clear that efficiency plays an important role when determining MIMO capacity, particularly at low SNR values. With a SNR of 15 dB, 25% capacity loss can be expected when low efficient antennas are

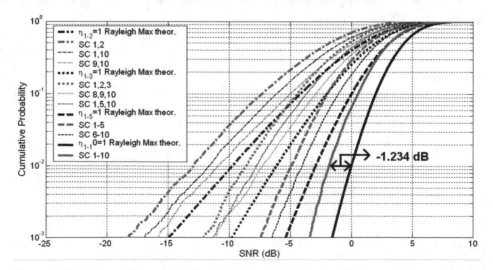

Figure 6.20   Cumulative probability distribution functions for different systems.

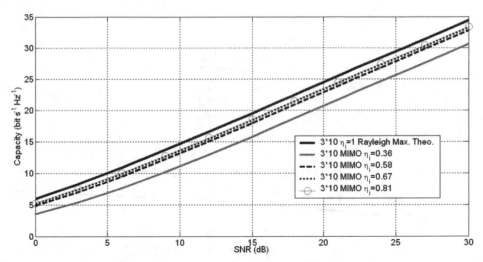

Figure 6.21   Measured MIMO capacity versus SNR for different total embedded radiation efficiency.

employed instead of high efficient antennas, or simply when the user is present in close proximity to the MIMO receiving antennas. Consequently, the combining possibilities and antenna topology for MIMO systems in the presence of the user acquires great importance.

In this sense, Figures 6.23 and 6.24 reproduce measured results of capacity loss due to efficiency for different $3 \times n$ MIMO systems wherein all branches have the same 0.85 and 0.46 radiation efficiencies, respectively. While there is a capacity loss in both figures, the effect of efficiency on MIMO capacity is clearly appreciated when the receiving antennas exhibit low efficiency values. Likewise, it is also observed from both figures that the additional capacity loss obtained when $R>T$ and employed receiving antennas are inefficient is mitigated when high SNRs

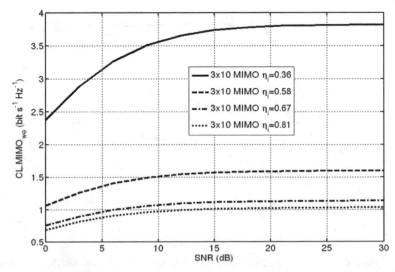

Figure 6.22   Measured capacity loss due to efficiency versus SNR for diverse total embedded radiation efficiency for 3x10 MIMO systems.

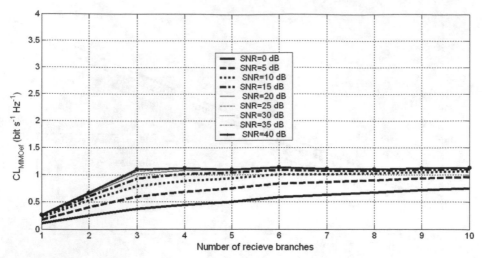

Figure 6.23 Measured capacity loss due to efficiency for MIMO systems with T= 3 and R as parameter versus SNR with η_i=0.85.

values are obtained. Similarly, the effect is more important when $R \leq T$, with a high slope that stabilizes when $R=T$ only for high SNRs. On the other hand, a quasi-linear increase with increasing R in MIMO capacity loss due to efficiency is observed when SNR is low. Therefore, the effect of efficiency on MIMO capacity also acquires great importance for low efficient antennas and low SNRs.

Yet it is highly unlikely that all receiving antennas have the same radiation efficiency despite the limited volume available at the handset scenario, not only because of a different distance to the user for each antenna, but also because each antenna design and topology may exhibit different radiation properties, as it was demonstrated in [55]. Thus, the effect of having diverse efficiencies within the MIMO system has also been studied [49]. Figure 6.25 depicts measured MIMO

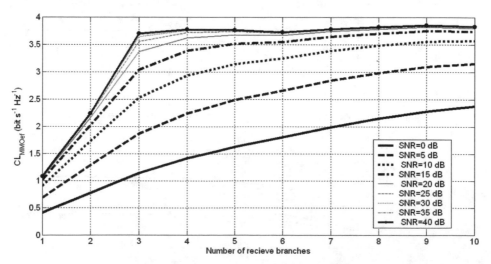

Figure 6.24 Measured capacity loss due to efficiency for MIMO systems with T= 3 and R as parameter for diverse SNRs with η_i=0.46.

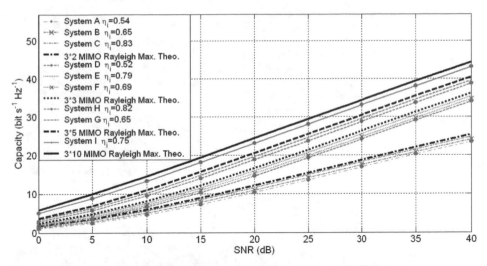

Figure 6.25 Measured MIMO capacity versus SNR for diverse MIMO systems.

capacity and capacity loss due to efficiency for different combining MIMO systems. Figures 6.26 and 6.27 illustrate the measured effect of adding a high efficiency receiving or transmitting antenna to 2×2 and 3×3 MIMO systems, respectively, both initially comprising low efficiency receiving antennas. As expected, the increment in MIMO capacity is clear when both transmitting and receiving antennas are simultaneously added to the system, even when these added antennas have a low efficiency. Yet, it is interesting to observe that, unlike what has been widely accepted for ideal combining antennas, there is a capacity increase beyond $R=T$ when high efficiency antennas are added to a low efficient MIMO system, which is more important as SNR increases. Adding a receiving antenna with an efficiency of 0.86 to a 3×3 MIMO system with $\eta_i=0.2$, for instance, provides the same MIMO capacity than a 4×4 MIMO system with $\eta_i=0.46$ at SNR=20 dB or a 4×4 MIMO system with $\eta_i=0.2$ at SNR=25 dB.

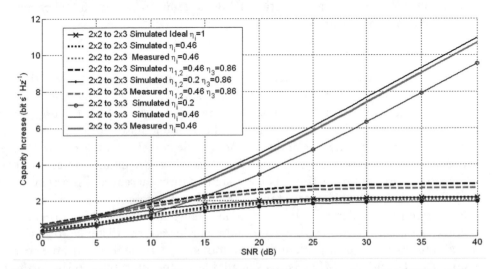

Figure 6.26 Measured capacity increase when adding antennas to a 2x2 MIMO system.

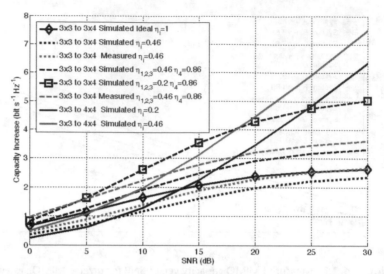

Figure 6.27 Measured capacity increase when adding antennas to a 3x3 MIMO system. (From: [49]. ©2008 IEEE. Reprinted with permission.)

This increment in capacity is also clearly observed from Figure 6.28, wherein the capacity loss due to efficiency is reduced when $R>T$; that is, when $\beta>1$, and when these added antennas have a high efficiency. It can also be concluded from Figure 6.28 that the capacity increases beyond $\beta=1$, or the equivalent reduction in capacity loss due to efficiency, can only be achieved when SNR>10 dB.

In summary, while efficiency plays a minor role on final effective diversity gain, it does affect the increasing diversity gain rate expected when increasing the number of receiving antennas. MIMO capacity loss due to efficiency, on the other hand, depends on the number of antennas, SNR, and efficiency in a complex way, acquiring great importance for low SNR values. Likewise, the full capacity of MIMO systems cannot be preserved with low efficient antennas, making the search for optimum and realistic MIMO antenna topology for each particular multipath scenario a difficult task.

6.3.2 Effect of User Presence

While the electromagnetic field-human body interaction has been thoroughly investigated in the literature, the effect of the presence of the user on MIMO performance has not received as much attention [21, 24, 58, 62, 64–66]. The presence of the user has demonstrated to have immediate influence on radiation patterns, input impedances and therefore on the correlation matrix, yet the effects are not fully understood and contradictory findings are commonplace. In [21] the envelope correlation coefficients were significantly increased and the mean effective gain (MEG) was decreased when the user was present. These changes showed a more important dependence to antenna orientation in [24]. In contrast, constant correlation coefficients have been found regardless of the distance from the receiving antenna to the user [62], and a simultaneous increment of both the correlation coefficients and the MEG is also available in the literature [66]. Furthermore, these contradictory results have also been found between simulated and measured

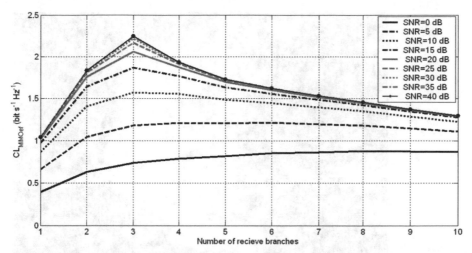

Figure 6.28 Measured capacity loss due to efficiency for MIMO systems with T=3 and R as a parameter.

results [65]. Contradictory findings can also be found for the effects on diversity gain and MIMO capacity [64, 65]. While most studies do not find significant changes in diversity performance when the user is present, measured diversity gain effects were found to be user- and antenna-dependent in [65]. Two different adult males and antennas were employed, and for one scenario the diversity gain was increased with one user and decreased with the other [65]. This effect was inverted when a different antenna type was employed [65]. Regarding the effects on MIMO capacity, a reduction in the MEG and a significant change in MIMO capacity was attributed to a power imbalance created by the presence of the user in [64], somehow contradicting what was previously neglected in [65] but later evaluated in ~2 dB in [67]. Thus, the complex dynamic role of the presence of the user on diversity gain and MIMO capacity has been identified as one of the issues for further study [68].

The effects of the presence of the user on the envelope correlation coefficient, power absorbed, diversity gain, and MIMO capacity were deeply investigated in [69]. The systems under test were formed by three transmit antennas, two receive antennas, and the SAM head phantom by MCL, filled with CENELEC A2400 (2,400 MHz) or A900 (900 MHz) head simulating liquids (HSLs). Measurements were carried out at 900 and 2400 MHz and were performed with and without the phantom. The three transmit antennas were orthogonal to each other and fixed to the chamber walls. The commercial Bluetest 001-B-019 $\lambda/2$ dipoles were used as the receive antennas. The measurement setup within the reverberation chamber is illustrated in Figure 6.29. In order to evaluate the solo-influence of the presence of the user on diversity gain and MIMO capacity, each isolated antenna was measured in a different position within the chamber loaded with the head phantom; that is, using uncorrelated branches. This avoided merging the effects due to the presence of the user to other effects such as mutual coupling, and it is similar to the switched-array technique for outdoor measurements [60]. Four different measurement scenarios, illustrated in Figure 6.30, were prepared using the setup depicted in Figure 6.29. In all scenarios dipole 1 was tilted according to the CEN-

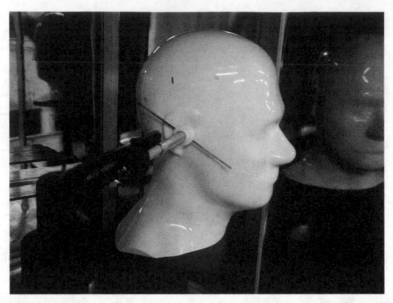

Figure 6.29   Measurement setup inside the reverberation chamber. (From: [69]. ©2007 IEEE.
Reprinted with permission.)

ELEC measuring procedure [70]. In scenario I, dipole 2 was placed in a position
orthogonal to dipole 1. Scenario II was prepared to distinguish the effect of the
user presence due to differences on the correlation coefficients from that due to
the power absorbed. In scenario II dipole 2 was rotated 90° relative to the position of dipole 1 with 10° steps. In this way different correlation coefficients were
obtained for the same user-antenna and dipole 1–dipole 2 separating distances.
Since a combination of both spatial and polarization diversity has proven to be an
effective way to improve diversity in reduced volumes such as the handset [46],
two more measurement scenarios were prepared. In scenario III (IV) the distance
from a 30°-rotated (90°-rotated) dipole 2 to the phantom was varied in 0.1λ
steps. In scenarios III and IV dipole 2 was always colocated to dipole 1, therefore
a minimum variation if the correlation coefficients were expected. Selection combining (SC) diversity gain measured results at 900 and 2,400 MHz in scenario I
can be observed from Figures 6.31 and 6.32, respectively. The presence of the user

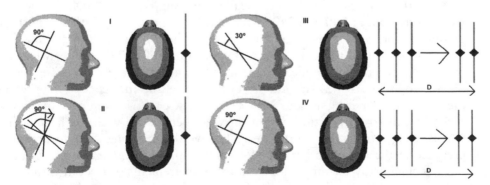

Figure 6.30   Measured scenarios. (From: [69]. ©2007 IEEE. Reprinted with permission.)

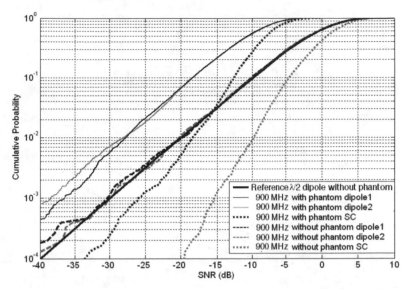

Figure 6.31   Measured cumulative probability density functions for single branch and SC diversity with and without the phantom at 900 MHz. (From: [69]. ©2007 IEEE. Reprinted with permission.)

clearly causes a displacement to the left of the CDF curves in all cases, which is proportional to the absorbed power. There is also a frequency-dependence of the effect of the user's presence, as also outlined in [64]. Since at 2,400 MHz more power is absorbed in the head than at 900 MHz, diversity gain loss is larger.

Table 6.2 summarizes measured SC diversity gain at 1% probability level, the envelope correlation coefficients between the two receive antennas, and the power

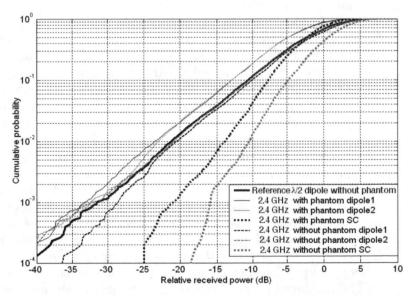

Figure 6.32   Measured cumulative probability density functions for single branch and SC diversity with and without the phantom at 2,400 MHz. (From: [69]. ©2007 IEEE. Reprinted with permission.)

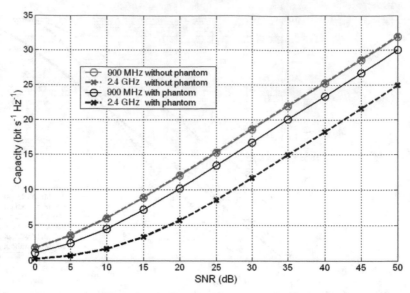

Figure 6.33   Measured MIMO capacity versus SNR. (From: [69]. ©2007 IEEE. Reprinted with permission.)

absorbed by the phantom. An interesting effect is observed from this table. While the envelope correlation coefficients are slightly reduced when the user is present, a relatively large diversity gain loss is accounted for. This indicates that the diversity gain loss has a stronger dependence on power absorbed than on the envelope correlation coefficient. Power absorbed has demonstrated to be strongly dependent on both antenna and human body topologies [71], and therefore that the effect of the user presence on diversity gain cannot be generalized, requiring specific and detailed studies.

MIMO capacity measured results with and without the phantom in scenario I can be observed from Figure 6.33. The presence of the user causes a loss in MIMO capacity, which is again proportional to absorbed power. A larger MIMO capacity loss can be seen at 2,400 MHz respect to that at 900 MHz. This difference demonstrates again the frequency-dependence of the effect of the user's presence, this time also for MIMO capacity. In a similar way to what happened to SC diversity gain, the slope of MIMO capacity curves is not altered by the presence of the user, except for low SNRs. A parallelism can be established between the rationale for the relationship between correlation, SNR and MIMO capacity of [72], and that of power absorbed in the user, SNR, and MIMO capacity. An increase in correlation implies a decrease in received SNR and consequently on MIMO capacity. Similarly, an increase in power absorbed by the user implies a decrease in received SNR and consequently on MIMO capacity. This rationale is also valid for SC diversity gain. Table 6.3 summarizes measured results in scenarios II and III. The results do not show an important role of the presence of the user on the envelope correlation coefficient, with a maximum change around 11%.

The combined effect of correlation and power absorbed due to the presence of the user on the CDF at 900 MHz in scenario III is depicted in Figure 6.34. The displacements in the CDF curve due to power absorbed and due to the envelope correlation coefficient are clearly distinguished. A larger CDF displacement effect

Table 6.2 A Comparison of Diversity Gain Measured Results

Frequency (MHz)	Presence of User	Diversity Gain (dB)	Envelope Correlation Coefficient	Power Absorbed (mW)
900	NO	9.51	0.139	-
900	YES	1.02	0.117	7.80
2400	NO	9.68	0.145	-
2400	YES	6.08	0.101	4.42

Table 6.3 Measured Results for Scenarios II and III at 900 MHz

II	Angular separation (°)	20	50	80
	Correlation coefficient with user	0.846	0.301	0.048
	Correlation coefficient without user	0.859	0.352	0.019
	Power absorbed (mW)	7.76	7.43	7.27
III	Distance to phantom	0	0.1 λ	0.2 λ
	Correlation coefficient with user	0.640	0.670	0.563
	Correlation coefficient without user	0.687	0.687	0.687
	Power absorbed (mW)	7.80	4.50	0.90

due to power absorbed is observed respect to the displacement due to the alteration in the envelope correlation coefficient when the user is present. This can also be observed by studying the effect of the presence of the user on MEG, depicted in Figure 6.35 for scenario IV. In this figure the reference antenna for the best branch MEG is always a dipole antenna without user [67]. The presence of the user clearly reduces MEG and unbalances branch power (~1.7 dB). When the antenna is displaced away from the user, its MEG tends to approach that of the best branch MEG and branch power tends to be balanced. In summary, the use of correlation coefficients is not appropriate for predicting MIMO capacity when in the presence of the user. This is a major drawback since correlation coefficients were typically employed to predict MIMO performance, and yet identifies new challenges for handset MIMO in practice.

In fact, the power absorbed in the user plays a more important role for MIMO capacity than the change on the correlation coefficients due to the user's presence for handset MIMO. Despite a reduction in the correlation coefficients when the user is present and in spite of the effects being frequency-dependent, a reduction in MEG, diversity gain, and MIMO capacity is expected when the user is present. Not only are the correlation coefficients not enough for properly predicting MIMO performance in the presence of the user, but also the expected degradation in MIMO performance due to the presence of the user is strongly

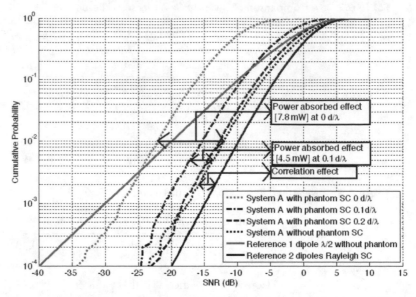

Figure 6.34 Measured cumulative probability density functions for system A. (From: [69]. ©2007 IEEE. Reprinted with permission.)

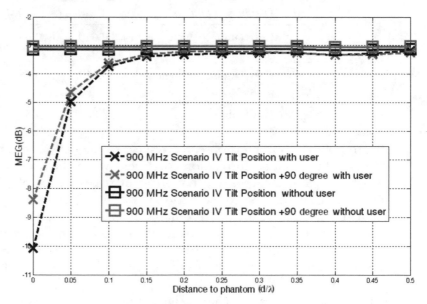

Figure 6.35 Measured MEG in scenario IV. (From: [69]. ©2007 IEEE. Reprinted with permission.)

dependent on both antenna topology and user characteristics. This calls for detailed studies including different users, antennas, and fading environments. Handset MIMO will not probably be straightforward and specific engineering efforts will have to be dedicated. Since different effects were observed in the correlation coefficients and MEG, the effect of the user's presence on antenna correlation for handset MIMO is also important.

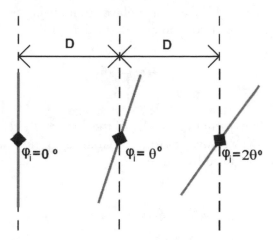

Figure 6.36   Geometry of the proposed antenna correlation function in [48]. (From: [48]. ©2007 IEEE. Reprinted with permission.)

6.3.3   Effect of Antenna Correlation

The correlation between different MIMO channels can be used to estimate the performance of the MIMO system since it is the result of interaction of the scattering environment and antenna properties. Losses due to correlation are quantified in [73]. Mutual coupling effects on the correlation coefficients are studied in [74] and 3-D models for obtaining the correlation coefficients are described in [75]. In many studies the correlation factor is employed to study MIMO system performance by assuming that the correlation among receive antennas is independent to the correlation between transmit antennas; that is, with no transmit-receive cross-correlation. This simplification makes a separable and analytically friendly model. When the antennas have the same radiation patterns, this model is known as the Kronecker model. When the correlation occurs only at either the receiver or transmitter ends, but not at both, the model is known as semicorrelated [14]. Some other models with cross-correlation assumptions are also available [76], including some analyses of the capacity and symbol error rate (SER) errors [77]. A novel function used to estimate both the spatial and angular correlation between two linearly-polarized antennas has recently been made available [48]. The new function is capable of estimating the correlation between two linearly polarized antennas that are spatially separated by a distance d and angularly separated by a rotation angle φ, as depicted in Figure 6.36. This is clearly useful for handset MIMO.

The spatial correlation function for dipole antennas in a multipath environment was originally developed by [78] and later validated using the reverberation chamber for isotropic environments [79] by

$$\zeta = \sin\left\lfloor z_{spatial} \right\rfloor / \left\lfloor z_{spatial} \right\rfloor \tag{6.16}$$

where $z_{spatial} = 2 * \pi * d_{spatial} / \lambda$ and $d_{spatial}$ is the interelement distance. This function is only valid for cross-polarization ratios XPR=1; that is, for conventional orthogonal polarization diversity. In a true polarization diversity (TPD) scheme the angular difference between two dipoles is defined as [48]

$$\varphi_{i-j} = \varphi_i - \varphi_j \qquad (6.17)$$

where φ_i and φ_j are the orientation angles of the dipoles. By representing an angular separation in TPD as equivalent to the spatial separation distance, a new correlation function was defined by

$$\zeta = \sin\lfloor z_{angular} \rfloor / \lfloor z_{angular} \rfloor \qquad (6.18)$$

where $z_{angular} = 2 * \pi * d_{angular}$ The angular separation $d_{angular}$ for Rayleigh fading scenarios with isotropic scattering was made equivalent to a spatial separation by

$$d_{angular} = \varphi'_{i-j} / 180 \qquad (6.19)$$

In this model the angular difference between two consecutive dipoles was transformed to an angle of the first quadrant in the following way

$$\varphi'_{i-j} = \begin{cases} abs(\varphi_{i-j}) & for \quad abs(\varphi_{i-j}) \leq 90 \\ 180 - abs(\varphi_{i-j}) & for \quad 90 \leq abs(\varphi_{i-j}) \leq 270 \\ 360 - abs(\varphi_{i-j}) & for \quad 270 \leq abs(\varphi_{i-j}) \leq 360 \end{cases} \qquad (6.20)$$

The model is only valid when the angular displacement between dipoles exists only in one axis. The complex correlation coefficients between multiaxis true polarization diversity recently proposed for three monopoles in a handset [42] is yet to be modeled. Both simulated and measured coefficients of the new correlation function are depicted in Figure 6.37. When using dipoles, there is a periodic pattern and a symmetry within the period of the correlation function, so that only 0° to 90° angular differences need to be simulated. The full potential of the novel correlation function can be exploited when combined to other spatial correlation functions.

In [48], two different ways are proposed and validated to combine spatial and angular diversity models. In the linear combined model the spatial distance is merged with its equivalent distance for angular diversity in a linear manner by

$$\zeta = \sin\lfloor z_{total} \rfloor / \lfloor z_{total} \rfloor \qquad (6.21)$$

where $Z_{total} = Z_{spatial} + Z_{angular}$ In Figure 6.38 the correlation for the linear combined model is depicted with spatial distances ranging from 0.1/λ to 0.5d/λ and

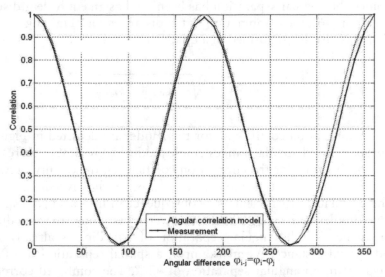

Figure 6.37   Simulated and measured angular correlation model versus angular difference of two dipoles. (From: [48]. ©2007 IEEE. Reprinted with permission.)

equivalent angular distances between 15° to 90°. As it happened to the diversity gain in [46], it is clear that when there is a large spatial separation, true polarization diversity can hardly reduce the already low correlation coefficient. In contrast, and also in a similar way to what happened for diversity gain in [46], when short spatial distances are employed, true polarization diversity can be very useful for reducing the high correlation coefficients. With just two dipoles the best option is to employ orthogonal polarizations; that is, an angular difference of 90° between them.

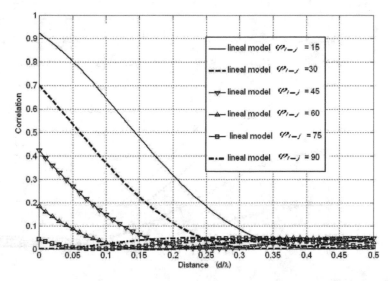

Figure 6.38   Linear model for the combined correlation function versus angular difference with spatial separation as a parameter. (From: [48]. ©2007 IEEE. Reprinted with permission.)

Since the angular separation has been used as an equivalent distance to spatial separation [46], the combination of the distances for obtaining z_{total} to be used in (6.21) was also be performed in a quadratic way by

$$z_{total} = \sqrt{z^2_{spatial} + z^2_{angular}}$$ (6.22)

The correlation coefficients for this model are depicted in Figure 6.39 for the same spatial and angular ranges employed in Figure 6.38. Figures 6.38 and 6.39 confirm, this time for the correlation coefficients, the duality between true angular and spatial diversity demonstrated in [46] for the diversity gain. The reduction in the correlation coefficient as a consequence of increasing the spatial separation of two dipoles with true polarization diversity is similar to the reduction obtained by incrementing the angular separation of two dipoles already separated by a fixed spatial distance. As an example, a spatial separation of ~0.2λ was found equivalent to an angular separation of ~36°. The combined correlation function models are useful for deriving true polarization diversity figures when combined with spatial diversity techniques. A comparison between measurements and the simulations using both the linear and quadratic combined models is illustrated in Figures 6.40 and 6.41, wherein the encountered differences are reproduced versus angular and spatial separation distances, respectively.

A careful observation of Figures 6.40 and 6.41 reveals that the linear combined model does not work well when there is a variation in the angular separation. Yet, while from Figure 6.40 it seems that the quadratic model has a moderate and quasi-constant error respect to a variation in the angular separation, from Figure 6.41 one can also observe that this behavior is not maintained when the spatial separation is altered. In fact, the linear- and quadratic-combined

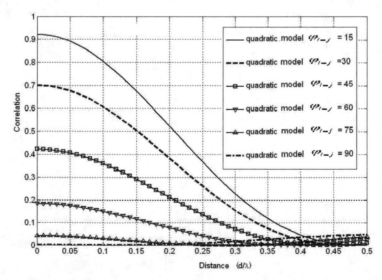

Figure 6.39 Quadratic model for the combined correlation function versus angular difference with spatial separation as a parameter. (From: [48]. ©2007 IEEE. Reprinted with permission.)

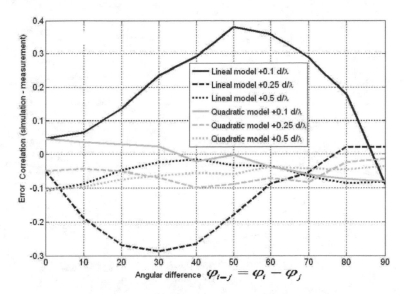

Figure 6.40   Errors in combined models for the correlation functions versus angular difference with distance as a parameter. (From: [48]. ©2007 IEEE. Reprinted with permission.)

models seem to follow opposite error behaviors when the spatial separation is altered. In both Figures 6.40 and 6.41 the maximum error in the quadratic model (0.10) is lower than the maximum error of the linear model (0.22). These errors are similar to other previously reported spatial-only models [75]. The differences, however, may put designed performance fulfillment in jeopardy, particularly for handset MIMO, and detailed correlation prediction is essential. Consequently, it may be possible that more complex algorithms and models could be obtained for a more accurate prediction of the correlation coefficients when using TPD or hybrid diversity schemes. With these recent developments, however, handset MIMO has revealed as a uniquely challenging task. Since in practice it is highly likely that a combination of techniques will achieve optimum performance, the diverse real effects (correlation, efficiency, presence of the user, TPD) will have to be accounted for early in the handset MIMO design process.

6.4   Combined Diversity Techniques for Optimum Handset MIMO Performance

The measured data obtained for TPD suggested that optimum performance for a reduced available volume would be achieved when combined to spatial diversity. Measured diversity gain of combined spatial-TPD 3×2 MIMO schemes are shown in Figures 6.42 and 6.43. In these figures dipoles are thus separated by both an angular and a spatial distance.

As expected, the combination of both spatial and true polarization diversity provided increased diversity gain with only two elements in the array. The combination has a stronger effect when both separations are not large. That is, when the spatial separation is large ($D \geq 0.24$), the angular separation can hardly improve the diversity gain, and vice versa when the angular separation is large ($d\theta \geq 54°$), the spatial separation can barely improve the diversity gain.

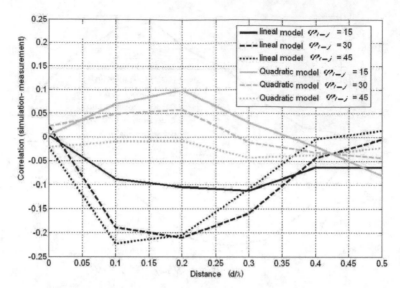

Figure 6.41   Errors in combined models for the correlation functions versus spatial separation with angular distance as a parameter. (From: [48]. ©2007 IEEE. Reprinted with permission.)

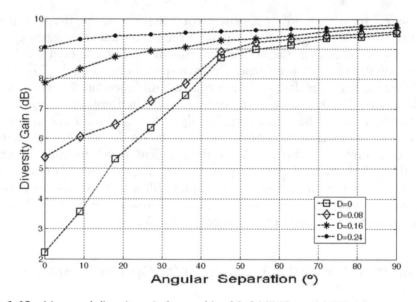

Figure 6.42   Measured diversity gain for combined 3x2 MIMO spatial TPD schemes.

This suggests that a good combination of the two values represents the most efficient technique for optimum diversity performance for the same reduced available volume for a complete array. Measurements for different linear and circular 3×6 combined-diversity MIMO systems were also performed. The diverse scenarios are listed in Table 6.4 and illustrated in Figure 6.44. For circular arrays the antenna separation D is taken as the array radius. The enormous potential of true polarization diversity for combined diversity schemes was confirmed by measuring the correlation coefficients, illustrated in Figure 6.45. The correlation coefficients depicted in Figure 6.45 for diverse array elements are measured respect to the first dipole in the array.

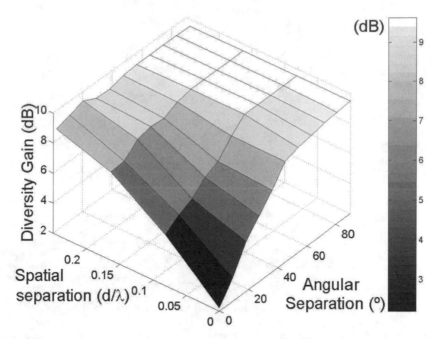

Figure 6.43   Measured diversity gain for combined 3x2 MIMO spatial TPD schemes.

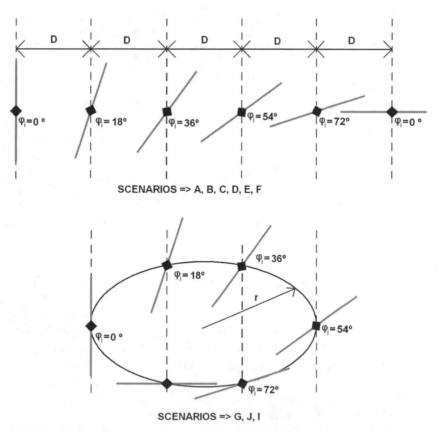

Figure 6.44   Analyzed geometries for combined spatial TPD schemes.

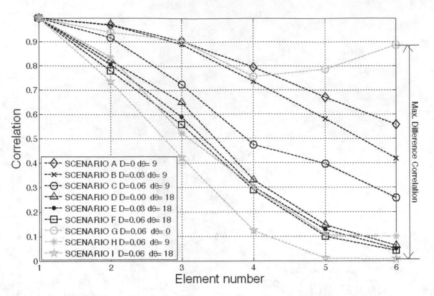

Figure 6.45   Measured correlation coefficients for the scenarios in Table 6.4.

Table 6.4   Combined Spatial TPD Schemes

	Antenna Separation D (d/λ)	Polarization Rotation angle dθ (°)	Array Type
Scenario A	0.00	9	Linear
Scenario B	0.03	9	Linear
Scenario C	0.06	9	Linear
Scenario D	0.00	18	Linear
Scenario E	0.03	18	Linear
Scenario F	0.06	18	Linear
Scenario G	0.06	0	Circular
Scenario H	0.06	9	Circular
Scenario I	0.06	18	Circular

From the correlation coefficients, MIMO capacity and diversity gain performance can be calculated for the MIMO array formed by the three wall-mounted
transmission antennas and the combination of up to six receiving dipole antennas.
This gives one possible 3×6 MIMO system, six different 3×5 MIMO systems, 15
different 3×4 MIMO systems, 20 different 3×3 MIMO systems, 15 different 3×2
MIMO systems, and six different 3×1 MIMO systems. For each MIMO system all
angular and spatial separation combinations can be employed, leading to a total
of 378 different measured MIMO systems. Figure 6.46 depicts the measured 3×3
MIMO capacity for different combined systems. Simulated 3×3 MIMO capacity
for uncorrelated Rayleigh fading channels is also depicted in Figure 6.46 for comparison purposes. When the angular or the spatial separations are large (i.e., the

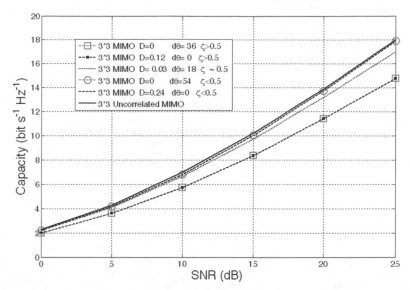

Figure 6.46 Simulated uncorrelated and measured 3x3 MIMO capacity (bps/s/Hz) for scenarios A to F.

correlation coefficients are small), the MIMO system perform at nearly full capacity. Yet when the spatial or the angular distances provide correlation coefficients over 0.5, a considerable capacity reduction is observed for single (either spatial or true polarization) diversity techniques. It is however interesting to observe that a combination of highly correlated both spatial and true polarization techniques (i.e., with low spatial and angular separations) also performs at nearly full capacity. The same excellent capacity results for combined diversity were obtained for circular arrays, as it is illustrated in Figure 6.47 for diverse 3×6 MIMO systems. It is interesting to note that more than three receiving antennas were employed in the tested systems of Figure 6.47. This combination is simply not possible with conventional orthogonal polarization diversity. Yet a hybrid system with three vertically polarized and three horizontally polarized antennas in the circular array would be possible. Such a system has not been reported, and it is also depicted in Figure 6.47 for comparison purposes. At SNR=10 dB, the combination of spatial and true polarization diversity nearly doubles the MIMO capacity of the spatial-only system. It is clear from this figure that true polarization diversity outperforms conventional orthogonal polarization diversity when more than two antennas are employed in reception.

The effect of adding or subtracting antennas to the MIMO system of scenario G is illustrated in Figure 6.48, wherein six MIMO subsystems all with $T=3$ but with a different number of receiving antennas number (1 to 6) are analyzed. As expected, diversity gain increases with increasing number of receiving antennas. The improvement is not so important when the number of antennas is large, particularly when the number of receiving antennas is larger than that of the transmitting antennas ($T=3$). To illustrate the effect of true polarization diversity as a potential way to increase diversity gain for volume-limited MIMO systems where spatial diversity is already employed, Figure 6.49 depicts the cumulative probability distribution function (CDF) of measured values so that diversity gain is obtained for three different subsystems all with $d\theta=9°$ but with $T=3$ and $R=6$ (all

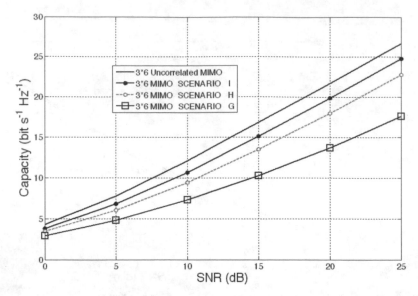

Figure 6.47   Simulated uncorrelated and measured 3x6 MIMO capacity for scenarios G to I.

antennas are used); $T=3$ and $R=3$ (only antennas 1, 3, and 6 are used), and $T=3$ and $R=2$ (only antennas 1 and 6 are used). Measured diversity gain is very similar for the three different MIMO subsystems of Figure 6.49, with maximum differences at 1% of the CDF of 1.72 dB. Should MIMO capacity depend solely on diversity gain, two antennas in reception for combined MIMO systems with $T=3$ would be the best choice (subsystem c), allowing the engineer to save space and processing efforts required for larger arrays. This is due to the fact that MIMO capacity strongly depends on the number of antennas with low correlation coefficients at both the receiving and transmitting ends, and adding antennas with high correlation coefficient does not really improve MIMO capacity. Notwithstanding the above, TPD has proven to be an ideal candidate for handset MIMO.

In conclusion, despite current 3G mobile communication systems already offering considerable improvement in data rates and throughput over their 2G counterparts, propagation, interference, and an ever-increasing demand for high data bit rates have prompted new challenges for antenna, network, and system design. This is particularly important in the downlink, and diversity reception in wireless system would help alleviating the problem. There are, however, inherent problems to handset MIMO in practice that are different from those encountered at conventional MIMO techniques. Consequently, widespread use of handset MIMO has yet to overcome the increase complexity of handsets and signal processing routines. Since 3G systems will clearly benefit from a reduction in the interference levels, and this could happen with just some smart handset MIMO devices, the future of handset MIMO is technology-dependent. Widespread use of 3G systems will undoubtedly lead to handset MIMO. Combining handset MIMO and currently used UMTS HSDPA can provide an extra 7.2 Mbps per 5-MHz bandwidth UMTS channel [80], which may increase sector throughput considerably [81].

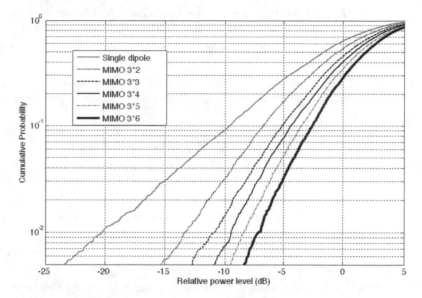

Figure 6.48   Cumulative probability distribution function (CDF) of measured diversity gain values for different combined MIMO systems.

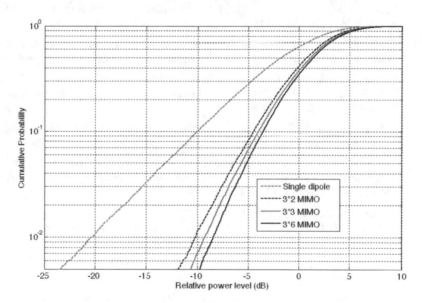

Figure 6.49   Cumulative probability distribution function (CDF) of measured diversity gain values for different combined MIMO systems.

References

[1]   Maci, S., et al., "Dual Frequency Patch Antennas," *IEEE Antennas and Propagation Magazine*, Vol. 39, No. 6, December 1997, pp. 13–20.

[2]   Sánchez-Hernández, D., and I. D. Robertson, "Analysis and Design of a Dual-Band Circularly Polarized Antenna," *IEEE Transactions on Antennas and Propagation*, Vol. 43, No. 2, February 1995, pp. 201–205.

[3]   Kossiavas, G., et al., "The C-Patch: A Small Microstrip Element," *Electronics Letters*, Vol. 25, No. 4, February 1989, pp. 253–254.

[4]   Huynh, T., and K.- F. Lee, "Single-Layer Single-Patch Wideband Microstrip Antenna," *Electronics Letters*, Vol. 31, No. 16, August 1995, pp. 1310–1312.

[5]   Carrerè, J. M., et al., Small Frequency Agile Antennas, *Electronics Letters*, Vol. 37, No. 12, June 2001, pp. 728–729.

[6]   Yang, F., et al., "Wide-Band E-Shaped Patch Antennas for Wireless Communications," *IEEE Transactions on Antennas and Propagation*, Vol. 49, No. 7, July 2001, pp. 1094–1100.

[7]   Turkoglu, O., et al., "E Shaped Antennas (ESA)," *Proc. IEEE Antennas and Propagation Society Int. Symposium*, Vol. 3, June 2002, pp. 504–507.

[8]   Guo, Y.- X., et al., "Compact Internal Multi-Band Antenna for Mobile Handsets," *IEEE Antennas and Wireless Propagation Letters*, Vol. 2, 2003, pp. 143–146.

[9]   Guo, Y.- X., et al., "Miniature Built-In Multi-Band Antennas for Mobile Handsets," *IEEE Transactions on Antennas and Propagation*, Vol. 52, No. 8, August 2004, pp. 1936–1944.

[10]   Deshmukh, A. A., and G. Kumar, "Compact Broadband S-Shaped Antennas," *Electronics Letters*, Vol. 42, No. 5, March 2006, pp. 260–261.

[11]   Boyle, K. R., and P.J. Massey, "Nine-Band Antenna System for Mobile Phones," *Electronics Letters*, Vol. 42, No. 5, March 2006, pp. 265–266.

[12]   Foschini, G. F., and M. Gans, "On Limits of Wireless Communication in a Fading Environment When Using Multiple Antennas," *Wireless Personal Communications*, March 1998, pp. 311–335.

[13]   Telatar, E., "Capacity of Multi-Antenna Gaussian Channels," *European Transactions on Telecommunications*, Vol. 10, No. 6, November/December 1999, pp. 585–595.

[14]   Kang, M., and M.- S. Alouni, "Capacity of Correlated MIMO Rayleigh Channels," *IEEE Transactions on Wireless Communications*, Vol. 5, January 2006, No. 1, pp. 143–155.

[15]   Kang, M., and M-S. Alouni, "Capacity of Correlated MIMO Rician Channels," *IEEE Transactions on Wireless Communications*, Vol. 5, No. 1, January 2006, pp. 112–120.

[16]   Svantesson, T., "Correlation and Channel Capacity of MIMO Systems Employing Multimode Antennas," *IEEE Transactions on Vehicular Technology*, Vol. 51, No. 6, November 2002, pp. 1304–1312.

[17]   Chiani, M., et al., "On the Capacity of Spatially Correlated MIMO Rayleigh-Fading Channels," *IEEE Transactions on Information Theory*, Vol. 49, No. 10, October 2003, pp. 2363–2371.

[18]   Goldsmith, A., et al., "Capacity Limits of MIMO Channels," *IEEE Journal on Selected Areas in Communications*, Vol. 21, No. 5, June 2003, pp. 684–702.

[19]   Hallbjörner, P., "The Significance of Radiation Efficiencies When Using S-Parameters to Calculate the Received Signal Correlation from Two Antennas," *IEEE Antennas and Wireless Propagation Letters*, Vol. 4, June 2005, pp. 97–99.

[20]   Martínez-Vázquez, M., et al., "Integrated Planar Multi-Band Antennas for Personal Communications Handsets," *IEEE Transactions on Antennas and Propagation*, Vol. 54, No. 2, February 2006, pp. 384–391.

[21]   Meksamoot, K., et al., "A Polarization Diversity PIFA on Portable Telephone and Human Body Effects on Its Performance," *IEICE Transactions on Communications*, Vol. E84.b, No. 9, September 2001, pp. 2460–2467.

[22]   Shin, H., and J. H. Lee, "Capacity of Multiple-Antenna Fading Channels: Spatial Fading Correlation, Double Scattering, and Keyhole," *IEEE Transactions on Information Theory*, Vol. 49, No. 10, October 2003, pp. 2636–2647.

[23]   Turkmani, A. M. D., et al., "An Experimental Evaluation of the Performance Two-Branch Space and Polarization Diversity Schemes at 1800 MHz," *IEEE Transactions on Vehicular Technology*, Vol. 44, No.2, May 1995, pp. 318–326.

[24]   Pedersen, G. F., and S. Skjaerris, "Influence on Antenna Diversity for a Handheld Phone by the Presence of a Person," *Proc. 47th IEEE Vehicular Technology Conference*, Vol. 3, May 1997, pp. 1768–1772.

[25]   Sarkar, T. K., et al., "Proper Interpretation of the Shannon Channel Capacity for the Vector Electromagnetic Problem," *Proc. Mediterranean Microwave Symposium*, 2006, pp. 531–534.

[26]   Getu, B. N., and J. B. Andersen, "The MIMO Cube—A Compact MIMO Antenna," *IEEE Transactions on Wireless Communications*, Vol. 4, No. 3, May 2005, pp. 1136–1141.

[27]   Chen, Z., et al., "Analysis of Transmit Antenna Selection/Maximal-Ratio Combining in Rayleigh Fading Channels," *IEEE Transactions on Vehicular Technology*, Vol. 45, No. 4, July 2005, pp. 1312–1321.

[28]   Wittneben, A., "A New Bandwidth Efficient Transmit Antenna Modulation Diversity Scheme for Linear Digital Modulation," *Proc. IEEE Int. Conference on Communications (ICC)*, Vol. 3, May 1993, pp. 1630–1634.

[29]   Molish, A .F., et al., "Capacity of MIMO Systems with Antenna Selection," *IEEE Transactions on Wireless Communications*, Vol. 4, No. 4, July 2005, pp. 1759–1772.

[30]   Greenstein, L. J., et al., "Moment-Method Estimation of the Ricean K-Factor," *IEEE Communications Letters*, Vol. 3, June 1999, pp. 175–176.

[31]   Tang, Z., and A. S. Mohan, "Characterize the Indoor Multipath Propagation for MIMO Communications," *Proc. Asia-Pacific Microwave Conference (APMC)*, Vol. 4, December 2005, p. 4.

[32]   Vaughan, R. G., "Switched Parasitic Elements for Antenna Diversity," *IEEE Transactions on Antennas and Propagation*, Vol. 47, No. 2, February 1999, pp. 399–405.

[33]   Aubrey, T., and P. White, "A Comparison of Switched Pattern Diversity Antennas," *Proc. 43rd IEEE Vehicular Technology Conference*, May 1993, pp. 89–92.

[34]   Mattheijssen, P., et al., "Antenna-Pattern Diversity Versus Space Diversity for Use at Handhelds," *IEEE Transactions on Vehicular Technology*, Vol. 53, No. 4, July 2004, pp. 1035–1042.

[35]   Dong, L., et al., "Multiple-Input Multiple-Output Wireless Communication Systems Using Antenna Pattern Diversity," *Proc. IEEE Global Telecommunications Conference (GLOBECOM)*, Vol. 1, November 2002, pp. 997–1001.

[36]   Lempiainen, J. J. A., and J. K. Laiho-Steffens, "The Performance of Polarization Diversity Schemes at a Base Station in Small/Micro Cells at 1800 MHz," *IEEE Transactions on Vehicular Technology*, Vol. 47, No. 3, August 1998, pp. 1087–1092.

[37]   Cho, K., et al., "Effectiveness of Four-Branch Height and Polarization Diversity Configuration for Street Microcell," *IEEE Transactions on Antennas and Propagation*, Vol. 6, No. 6, June 1998, pp. 776–781.

[38]   Andrews, M. A., et al., "Tripling the Capacity of Wireless Communications Using Electromagnetic Polarization," *Nature*, Vol. 409, January 2001, pp. 316–318.

[39]   Lukama, L., et al., "Three-Branch Orthogonal Polarization Diversity," *Electronics Letters*, Vol. 37, No. 20, September 2001, pp. 1258–1259.

[40]   Tumbuka, M. C., and D.J. Edwards, "Investigation of Tri-Polarised MIMO Technique," *Electronics Letters*, Vol. 41, No. 3, February 2005, pp. 137–138.

[41]   Das, N. K., et al., "An Experiment on MIMO System Having Three Orthogonal Polarization Diversity Branches in Multipath-Rich Environment," *Proc. 60th IEEE Vehicular Technology Conference*, Vol. 2, September 2004, pp. 1528–1532.

[42]   Dong, L., et al., "Simulation of MIMO Channel Capacity with Antenna Polarization Diversity," *IEEE Transactions on Wireless Communications*, Vol. 4, No. 4, July 2005, pp. 1869–1873.

[43]   Eggers, P. C. F., et al., "Antenna Systems for Base Station Diversity in Urban Small and Micro Cells," *IEEE Journal on Selected Areas in Communications*, Vol. 11, No. 7, September 1993, pp. 1046–1057.

[44]   Oestges, C., and A. J. Paulraj, "Beneficial Impact of Channel Correlations on MIMO Capacity," *Electronics Letters*, Vol. 40, No. 10, May 2004, pp. 606–608.

[45]   Marzetta, T. L., et al., "Fundamental Limitations on the Capacity of Wireless Links That Use Polarimetric Antenna Arrays," *Proc. Int. Symposium on Information Theory (ISIT 2002)*, July 2002, p. 51.

[46]   Valenzuela-Valdés, J. F., et al., "The Role of Polarization Diversity for MIMO Systems Under Rayleigh-Fading Environments," *IEEE Antennas and Wireless Propagation Letters*, Vol. 5, 2006, pp. 534–536.

[47]   Oestges, C., et al., "Propagation Modeling for MIMO Multipolarized Fixed Wireless Channels," *IEEE Transactions on Vehicular Technology*, Vol. 53, No. 3, May 2004, pp.644–654.

[48]   Valenzuela-Valdés, J. F., et al., "Estimating Combined Correlation Functions for Dipoles in Rayleigh-Fading Scenarios," *IEEE Antennas and Wireless Propagation Letters*, Vol. 6, 2007, pp. 349–352.

[49]   Valenzuela-Valdés, J. F., et al., "The Influence of Efficiency on Receive Diversity and MIMO Capacity for Rayleigh-Fading Channels," *Antennas and Propagation, IEEE Transactions*, Vol. 56, No. 5, May 2008, pp. 1444–1450.

[50]   Andersen, J. B., and B. N. Getu, "The MIMO Cube—A Compact MIMO Antenna," *Proc. 5th IEEE Int. Symposium on Wireless Personal Multimedia Communications*, Vol. 1, October 2002, pp. 112–114.

[51]   Getu, B. N., and, R. Janaswamy, "The Effect of Mutual Coupling on the Capacity of the MIMO Cube," *IEEE Antennas and Wireless Propagation Letters*, Vol. 4, 2005, pp. 240–244.

[52]   Chiu, C. Y., and R. D. Murch, "Experimental Results for a MIMO Cube," *Proc. IEEE Antennas and Propagation Society Int. Symposium*, July 2006, pp. 2533–2536.

[53]   Ying, Z., et al., "Diversity Antenna Terminal Evaluation," *Proc. IEEE Int. Symposium on Antennas and Propagation*, Vol. 2A, July 2005, pp. 375–378.

[54]   Winters, J. H., "The Diversity Gain of Transmit Diversity in Wireless Systems with Rayleigh Fading," *IEEE Transactions on Vehicular Technology*, Vol. 47, No. 1, February 2001, pp.119–123.

[55]   Karaboikis, M., et al., "Three Branch Antenna Diversity System on Wireless Devices Using Various Printed Monopoles," *Proc. IEEE Int. Symposium on Electromagnetic Compatibility*, Vol. 1, May 2003, pp. 135–138.

[56]   Lozano, A., and A. M. Tulino, "Capacity of Multiple-Transmit Multiple-Receive Antenna Architectures," *IEEE Transactions on Information Theory*, Vol. 48, No. 12, December 2002, pp. 3117–3127.

[57]   Loyka, S., et al., "Estimating MIMO System Performance Using the Correlation Matrix Approach," *IEEE Communications Letters*, Vol. 6, No. 1, January 2002, pp. 19–21.

[58] Ogawa, K., et al., "An Analysis of the Performance of a Handset Diversity Antenna Influenced by Head, Hand, and Shoulder Effects at 900 MHZ: Part I—Effective Gain Characteristics," *IEEE Transactions on Vehicular Technology*, Vol. 50, No. 3, May 2001, pp. 830–844.

[59] Tay, R. Y. S., et al., "Dipole Configurations with Strongly Improved Radiation Efficiency for Hand-Held Transceivers," *IEEE Transactions on Antennas and Propagation*, Vol. 46, No. 6, June 1998, pp. 798–806.

[60] Jensen, M. A., and J. W. Wallace, "A Review of Antennas and Propagation for MIMO Wireless Communications," *IEEE Transactions on Antennas and Propagation*, Vol. 52, No. 11, November 2004, pp. 2810–2824.

[61] Kildal, P. S., and C. Carlsson, "Detection of a Polarization Imbalance in Reverberation Chambers and How to Remove It by Polarization Stirring When Measuring Antenna Efficiencies," *Microwave and Optical Technology Letters*, Vol. 34, No. 2, July 2002, pp. 145–149.

[62] Ogawa, K., et al., "An Analysis of the Performance of a Handset Diversity Antenna Influenced by Head, Hand, and Shoulder Effects at 900 MHz: Part II—Correlation Characteristics," *IEEE Transactions on Vehicular Technology*, Vol. 50, No.3, May 2001, pp. 845–853.

[63] Pedersen, G. F., and J .B. Andersen, "Handset Antennas for Mobile Communications: Integration Diversity, and Performance," in *Review of Radio Science 1996–1999*, W. R. Stone, (ed.), New York: Oxford University Press, 1999, pp. 119–137.

[64] Waldschmidt, C., et al., "MIMO Handheld Performance in Presence of a Person," *Proc. URSI Int. Symposium on Electromagnetic Theory*, 2004, pp. 81–83.

[65] Green, B. M., and M.A. Jensen, "Diversity Performance of Dual Antenna Handsets Near Operator Tissue," *IEEE Transactions on Antennas and Propagation*, Vol. 48, No. 7, July 2000, pp. 1017–1024.

[66] Kemp, D. C., and Y. Huang, "Antenna Diversity and User Interaction at 1800 MHz," *Proc. IEEE Antennas and Propagation Society Int. Symposium*, Vol. 3, July 2005, pp. 479–482.

[67] Nielsen, J. O., and G. F. Pedersen, "Mobile Handset Performance Evaluation Using Radiation Pattern Measurements," *IEEE Transactions on Antennas and Propagation*, Vol. 54, No. 7, July 2006, pp. 2154–2165.

[68] Kildal, P. S., and K. Rosengren, "Correlation and Capacity of MIMO Systems and Mutual Coupling, Radiation Efficiency, and Diversity Gain of Their Antennas: Simulations and Measurement in a Reverberation Chamber," *IEEE Communications Magazine*, Vol. 44, No. 12, December 2004, pp. 104–112.

[69] Valenzuela-Valdés, J. F., et al., "Effect of User Presence on Receive Diversity and MIMO Capacity for Rayleigh-Fading Channels," *IEEE Antennas and Wireless Propagation Letters*, Vol. 6, 2007, pp. 596–599.

[70] CENELEC EN 50361, *Basic Standard for the Measurement of Specific Absorption Rate Related to Human Exposure to Electromagnetic Fields from Mobile Phones (300 MHz–3 GHz)*, European Standard. Brussels. EN50361:2001 E.

[71] Christ, A., et al., "The Dependence of Electromagnetic Far-Field Absorption on Body Tissue Composition in the Frequency Range from 300 MHz to 6 GHz," *IEEE Transactions on Microwave Theory and Techniques*, Vol. 54, No. 5, May 2006, pp. 2188–2195.

[72] Loyka, S., "Channel Capacity of MIMO Architecture Using the Exponential Correlation Matrix," *IEEE Communications Letters*, Vol. 5, No. 9, May 2001, pp. 369–371.

[73]   Li, X., and Z. P., Nie, "Mutual Coupling Effects on the Performance of MIMO Wireless Channels," *IEEE Antennas and Wireless Propagation Letters*, Vol. 3, No. 9, 2004, pp. 344–347.

[74]   Derneryd, A., et al., "Signal Correlation Including Antenna Coupling," *Electronic Letters*, Vol. 40, No. 3, February 2004, pp. 157–159.

[75]   Hui, P., et al., "3D Autocorrelation Coefficients of Dipole Antenna," *Electronics Letters*, Vol. 42, No. 5, March 2006, pp. 257–258.

[76]   Tulino, A. M., et al., "Impact of Antenna Correlation on The Capacity of Multiantenna Channels," *IEEE Transactions on Information Theory*, Vol. 51, No. 7, July 2005, pp. 2491–2509.

[77]   Oestges, C., et al., "Impact of Fading Correlations on MIMO Communication Systems in Geometry-Based Statistical Channel Models," *IEEE Transactions on Wireless Communications*, Vol. 4, No. 3, May 2005, pp. 1112–1120.

[78]   Jackes, W. C., *Microwave Mobile Communications*, New York: Wiley, 1974.

[79]   Hill, D. A., and J. M. Ladbury, "Spatial/Correlation Functions of Fields and Energy Density in a Reverberation Chamber," *IEEE Transactions on Electromagnetic Compatibility*, Vol. 44, No. 1, February 2002, pp. 95–101.

[80]   3GPP, *Multiple Input Multiple Output (MIMO) Antennae in UTRA*, Technical Report TR 25.876 V7.0.0, 2007.

[81]   Zooghby, A. E., *Smart Antenna Engineering*, Norwood, MA: Artech House, 2005.

CHAPTER 7

Communication Performance of Mobile Devices

Gert Frølund Pedersen and Jesper Ødum Nielsen

7.1   Introduction

In the late 1990s, validation of handheld antennas was becoming very important as the antenna design changed from more traditional add-on whips and normal mode helicals to custom-made integrated antennas. The so-called mean effective gain (MEG) was introduced as a method of evaluating the performance of antennas in a mobile environment, and is defined as the ratio of the average power received by the antenna under test to the average power received by some reference antenna, while they are both moving in the same environment [1–49]. This method of evaluating antennas directly via measurements in the mobile environment should include the person using the handset for realistic performance results. The MEG is a very good measure of how well the antenna performs in a specific environment but often the measurements are carried out by using only a single "test person" or with a head phantom. This may explain why validation made in the 1990s by different mobile network operators of the same handset model have often led to very different performance assessment. It was then relevant to ask how much influence the mobile user has on the antenna performance.

To answer this question a measurement campaign was carried out and different antennas were evaluated in the same environments, with different persons. The measurements were carried out at 1,880 MHz having 200 test persons use the handsets in normal speaking position. The mock-up handheld consisted of a commercially available GSM-1800 handheld equipped with a retractable whip and a normal mode helical antenna that was modified to also include a back-mounted patch antenna. Two 50 Ω cables were used to connect the antennas to the receiving equipment. The mock-up therefore included three antennas connected to two connectors, a patch antenna on one connector and either the whip or the normal mode helix (when not retracted) on the other connector. The measurements were carried out by asking each test person to hold the handheld in a way he or she felt was a natural speaking position (see Figure 7.1). Then the person was asked to follow a path marked with tape on the floor. The path was a rectangle of some 2 × 4 meters and each record of data lasted 1 minute corresponding to three to four rounds. To record all three antennas each person had to follow the path for 1 minute, change the whip antenna to the helix and walk along the path once again. Hence, the measurements on the first trip were used to record the performance of the whip and the patch, and the measurements of the

Figure 7.1   One of the test persons holding the handheld in what he feels is a natural speaking position during measurement. (From: [21]. ©1998 IEEE. Reprinted with permission.)

second trip were used to record the performance of the helix antenna and the patch once again.

Altogether four locations were selected, one path on each floor of a four-storey building and 50 test persons were used on each floor. The windows on level 3, level 2, and the ground level were facing towards the transmitter but there was no line of sight due to higher buildings in between. On the first floor the windows were facing opposite the transmitter. The handset was connected to a dual-channel wide-band correlation sounder in order to record two antennas at a time. The carrier frequency was 1,890 MHz and a bandwidth of 20 MHz was used to suppress the fast fading. The instantaneous dynamic range of the sounder was 45 dB and the overall dynamic range was 80 dB with a linearity of 1 dB. To match a typical urban GSM-1800 microcell, the transmitter antenna was located approximately 700m away on the sixteenth level of a high building in an urban environment. The transmit antenna was a 60° sector antenna with a beamwidth of 5° in elevation and it was tilted mechanically some 4° down. Figure 7.2 shows the transmit antenna together with the view of the environment. The building in which the measurements were performed is hidden by other buildings in the picture and therefore no line of sight exists between the transmitter and the handheld receiver.

On each floor a measurement was made for every combination of test person and antenna type. Each measurement consisted of 1,000 impulse responses, from which the average power was computed as the performance measure. As a reference, all combinations of antennas and floors were also measured without a person, by mounting the handset on a wooden stick at a 60° tilt angle from vertical. These measurements were repeated three times to allow for checks of repeatability.

Figure 7.2 Picture of the transmitting antenna (the antenna is vertically mounted) and the view of the used urban area.

The results are shown summarized as mean and standard deviations in Table 7.1 and as an example all the results from the first floor are shown in Figure 7.3. From the results it is clear that the helix performs worse than the two other antennas, since it has lower averages and larger spread value. Peak differences of up to 10 dB can be found for the helix antenna, while for the whip and patch antennas peak differences of the order of 5 to 7 dB are found.

Having found that the variation in received power from one person to another can be very large, possible explanations were investigated. The first parameter to investigate is the repeatability (or uncertainty) of the measurement. In Figure 7.4 the received power in "free space" is shown for all antennas on floors one to three. Three repeated measurements were recorded in order to examine the repeatability. Both the mean and spread are plotted on the figure showing that the spread is only ±0.5 dB. Even when a person was present the spread is very small as can be found from the fact that the patch antenna was recorded twice for each person (see Table 7.1) for patch W and patch H. This confirmed that the results were repeatable and reliable. Also note that the power was higher in free space than if a person was present, especially for the helix.

The influence of the hand the test persons were using (left or right) was also investigated. Measurements with two persons were made on the same path (first floor) 20 times each, 10 times with the whip and patch (five with the left hand and five with the right hand) and 10 with the helix and patch. Figure 7.5 displays the results for person A and it is clear that there is a major difference between a left-handed and right-handed person. The difference is most significant for the patch antenna (some 5 dB) and less significant for the whip (some 1 dB). In Table 7.2 the results for both persons are shown and the results for person B are similar

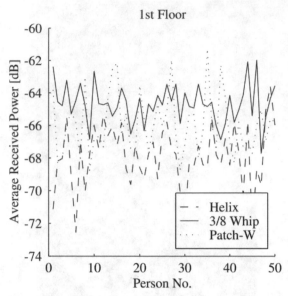

Figure 7.3   Received power by three antennas on a handheld for 50 different persons on the first floor of a building. (From: [21]. ©1998 IEEE. Reprinted with permission.)

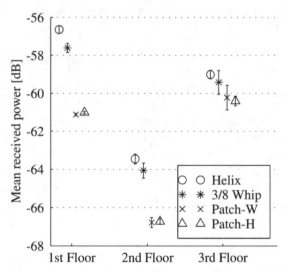

Figure 7.4   Received power when no person was present. Both average values and spread among three measurements are shown for each antenna and on each floor. (From: [21]. ©1998 IEEE. Reprinted with permission.)

to those for person A shown in Figure 7.5. Note that the spread is low among the five repeated measurements.

The results described above surprised the authors and naturally doubts were raised whether these remarkable results were due to the setup used, and thus would not be reproducible in an actual cellular network. For example, the measurements were made with a channel sounder where a special Tx unit was used which was obviously not exactly like a real base station, and at the Rx side the antennas in the handset were connected to the sounder with coaxial cables. This could influence the results, since the cables to some degree become part of the

Table 7.1   Averaged Received Power in dB and Standard Deviation for Each Antenna on Each Floor

Antenna	Ground Floor	First Floor	Second Floor	Third Floor
Whip	−78.7±1.3	−64.7±1.2	−71.1±1.3	−64.5±2.1
Patch W	−78.8±1.5	−65.9±1.9	−70.3±1.4	−62.7±1.7
Patch H	−78.8±1.5	−66.0±1.7	−70.3±1.3	−63.0±1.7
Helix	−81.7±1.6	−67.9±1.9	−73.6±1.8	−66.9±2.5

Patch W is the patch recorded at the same time as the whip and patch h is the same patch recorded at the same time as the helix. *Source:* [21].

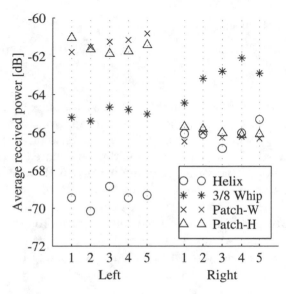

Figure 7.5   Received power in decibels for person A when using the left hand and right hand to hold the handset towards the head. The measurements were repeated five times to allow calculation of the spread. (From: [21]. ©1998 IEEE. Reprinted with permission.)

antenna, thus changing the effective radiation pattern and possibly change when the cables are flexed.

For these reasons it was decided to try to reproduce the results, but this time in a real network using operational GSM handsets. As part of the normal operation of a GSM network, a number of parameters are measured regularly by the base stations as well as by the handset during calls. For example, the received signal strength is measured. In the network this information is transferred from the base stations to other parts of the network over the so-called Abis interface. A way to characterize the overall performance of a handset in the operating network is to collect the information transferred on the Abis during calls and process the data to obtain, for example, the average receive power level. In this way the actual GSM network becomes the measurement system and the performance evaluation can be done when the handset is in normal use [2].

Table 7.2   Average Received Power in Decibels for Two Persons When Using the Left Hand (and Left Side of Head) and the Right Hand

	Person	
Antenna	A	B
Whip left	-65.0±0.3	-64.3±0.4
Whip right	-63.1±0.9	-63.2±0.3
Patch H left	-61.5±0.3	-60.8±0.2
Patch W left	-61.3±0.4	-60.7±0.2
Patch H right	-66.0±0.2	-66.2±0.4
Patch W right	-66.3±0.2	-65.3±0.3
Helix left	-69.4±0.5	-68.2±0.4
Helix right	-66.1±0.5	-64.4±0.4

Source: [21].

The measurements were carried out in a building at Aalborg University, Denmark, where a small base station is located approximately 75m away. The base station serves both GSM-900 and GSM-1800 cells, and data from both frequency bands were used. The measurement building is situated in the outskirts of the city and is a new four-story office building mainly made of reinforced concrete with an outer brick wall. Most inner walls are made of plasterboard. The measurements took place in corridors of the basement and the first floor. On the first floor the corridor provides access to small offices on both sides, where the offices on one side have windows towards the base station. The measurements involved 12 users, each testing four different GSM handsets using both the left and the right hand/side of the head. The handsets are shown in Figure 7.6.

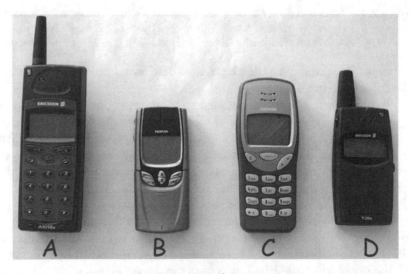

Figure 7.6   The handsets used for the in-network measurements.

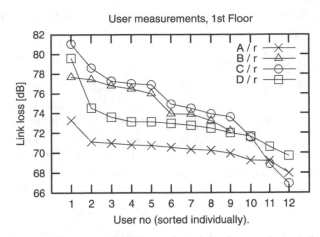

Figure 7.7   The link loss measured with different users for the four handsets on the first floor. The right hand is used in all cases. (From: [3]. ©2000 IEEE. Reprinted with permission.)

As an example, the mean link loss for these handsets is shown in Figure 7.7 for right-hand usage on the first floor. In the figure it is noticed that the variation in link loss depends on the handset. For example, handset A had a maximum difference of about 5 dB with most values within about 3 dB, whereas for handset C the maximum difference was about 14 dB. The results illustrate that the previous results found using channel sounding and the mock-up handset are realistic. The loss in the link budget introduced by the user, called the bodyloss, is defined as the ratio of received power by the handset with the person present to the power received in the absence of the person. Figure 7.7 also shows large differences in performance between the four handsets, and thus optimization of the design with respect to bodyloss is advisable. These results were also in line with what was felt by operators, that the performance of handsets in some cases was too low and that some standardized testing was needed. More results and details of the measurements can be found in [3].

7.2   Antenna Efficiency

Often the term antenna efficiency is used when referring to how good the antenna is at either receiving the power available in the air or to radiate the power available from the transmitter. But the term antenna efficiency may or may not include certain losses connected to the transfer of the energy from, for example, the transmitter to the air (i.e., impedance matching, polarization matching, or load-pull by the power amplifier). This section starts with the definition of the antenna efficiency and describes ways to obtain the antenna efficiency through measurements with particular focus on electrically small antennas on small devises. The section ends describing the practical measurements of "overall" antenna efficiency of small terminals like mobile phones, the so-called total radiated power (TRP) and total isotropic sensitivity (TIS) used in emerging standards. The "overall" antenna efficiency does include the matching loss and losses due to load-pull of transmitters but it does not include any polarization or radiation direction losses. To include such losses another term is needed, namely MEG. MEG will be described in more detail in Section 7.4.

7.2.1 Radiation Efficiency

The antenna radiation efficiency is defined as the ratio of the total radiated power to the power accepted by the antenna at the input. This means that the impedance mismatch is not part of the antenna radiation efficiency as defined by the IEEE standard definitions of terms for antennas [4]. The losses affecting the antenna radiation efficiency can be of both conductive and dielectric nature (i.e., losses in pure conductive materials as well as losses in lossy dielectric materials). The total antenna efficiency can be obtained by including the antenna mismatch loss (i.e., $1-|\Gamma|^2$, where Γ is the reflection coefficient).

7.2.2 How to Obtain the Antenna Efficiency

7.2.2.1 Temperature Increase Measurements

One very simple and direct way to obtain the radiation efficiency, at least in theory, is to connect the antenna to a generator that can deliver a relatively high power signal in the range of 1W and then measure the temperature increase or temperature when the steady state is reached. The antenna can then be replaced by a resistor connected to either the generator or just a power supply and the input power to the resistor is now varied until the same increase or steady-state temperature is reached. This method requires the ambient temperature to be constant and in practice it is not an easy way to measure antenna losses, at least not for small losses. Although it is difficult to measure the absolute loss this way, it is relatively easy to use a similar measure to find the reason or location of the losses. With the antenna connected to the high power generator a thermographical camera can be used to see where the power is absorbed. Special care needs to be given to other material characteristics such as the thermal conductivity and infrared reflections.

7.2.2.2 Wheeler Cap and Reverberation Chambers

The Wheeler cap method is a well-known method for obtaining the antenna efficiency and requires one measurement of the impedance with the antenna in an anechoic room or in free space and another where the entire antenna is surrounded by a conducting cap. Using these measurements and an equivalent parallel or series RLC circuit model it is possible to estimate the antenna efficiency [5]. Most electrically small antennas can be modeled in this way.

A drawback of this method is that the impedance needs to be measured and hence the method cannot take into account aspects such as change in the power delivered by the amplifier due to a loading different from nominal (load-pull). In an attempt to create a test of the entire device under test (DUT) in realistic environments, the so-called stirred mode chamber [6, 7] was created. When the DUT is placed inside this shielded chamber with metal walls, a number of moving metallic parts are used to create a changing field, and thereby simulate an environment. The main problem with these methods using artificial environments is in controlling the signal distribution inside the room and that the created mobile environment is not necessarily typical for a mobile in actual use [8]. Since the radiation pattern is not measured directly with this method it is not possible to test other environments using postmeasurement processing.

7.2.2.3   Radiation Pattern

The antenna efficiency can easily be obtained if the radiated intensity is known on a closed surface surrounding the antenna by integration of the intensity weighted by the distance to the antenna. In practice it is often implemented by measuring on a full sphere where both polarizations are measured on a grid of, say, 15°×15°. It is then possible to calculate the radiated power by simple integration and compare it to the power delivered to the antenna by the generator or transmitter to obtain the antenna efficiency. Further, it is possible to obtain the peak gain (maximum of the gain as a function of direction) and the coordinates for the peak gain that is often needed for type approval of a device. By using an external generator connected to the antenna under test (AUT), it is possible to make accurate measurements of the antenna efficiency due to the well-defined input signal. It should be noted that the radiated power calculated from the measurements of radiation patterns can only be compared to the measured input power to obtain the correct antenna efficiency under the assumption that the power delivered to the antenna from the power amplifier does not change due to an antenna impedance different from 50 Ω and the assumption that the antenna impedance can be measured accurately with a cable. It should also be noted that connecting an external conductive cable could disturb the radiation pattern if the AUT is a small unit. However, this will not always influence the antenna efficiency significantly. The issue of connecting a conductive cable and load-pull will be detailed further below.

7.2.3   Conductive Cable Problem

Connecting a conductive cable to a wireless device for the sole reason of testing the radiation performance can significantly change the radiation characteristics. If the radiation performance of the wireless device is sought for the case of no cable connected (i.e., for a device that typically operates without any cable connected), special care must be taken when connecting a conductive cable, for example, when supplying power to the device or signal to the antenna. Both the radiation pattern and the input impedance may change significantly if a conductive cable is connected.

To illustrate this problem some FDTD simulations were carried out on a mobile handset, modeled as shown in Figure 7.8. The simulations were done at three frequency bands centered about 450, 900, and 2,400 MHz. Three cases were simulated: (a) no cable was attached to the box (reference); (b) a straight cable was attached to the box; and (c) a cable was attached to the box, with a right angle bend 30 mm from the box.

The impedances for all combinations of frequency bands and cable setup are shown in Figure 7.9. Attaching a cable to the handset significantly changes the impedance, especially at the low frequencies. Figure 7.10 displays the changes in the radiation patterns, where attaching the cable again changes the results significantly. Only the θ-polarization is shown since the gain in the φ-polarization is negligible when no cable or a straight cable is attached. However, the cable actually

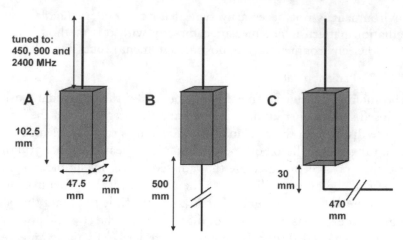

Figure 7.8   Dimensions of simulated mobile handset equipped with a monopole antenna tuned to three different frequencies. The three cases are for the cases (a) no cable attached, (b) a straight cable attached, and (c) a bent cable attached.

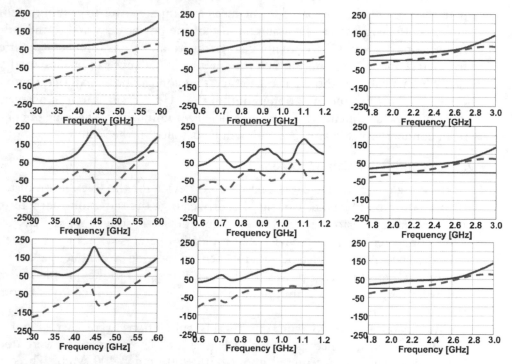

Figure 7.9   Simulated impedance in ohms for three frequency bands, 0.3–0.6, 0.6–1.2, and 1.8–3 GHz for no cable connected (top row), straight cable (middle row), and bent cable (bottom row). The real and imaginary part of the impedance are shown with solid and dashed lines, respectively.

changes the polarization properties, so that the ϕ-polarization becomes as shown in Figure 7.11 when a bent cable is attached.

From the examples shown it is evident that connecting extra conductive cables to small devices with the purpose of measuring, for example, the radiation pattern may be problematic. First, the cable acts as a radiation source, excited by current

Without cable connected

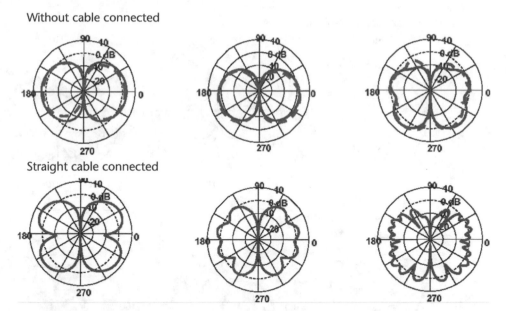

Straight cable connected

Figure 7.10 Simulated radiation pattern for the θ-polarization with (top row) and without (bottom row) a straight cable connected. The dashed line is for the similar measurements. Similar to Figure 7.9, the different columns represent the frequency bands, 0.3–0.6, 0.6–1.2, and 1.8–3 GHz. (From: [11]. ©2001 IEEE. Reprinted with permission.)

originating from the handset, and second, the cable acts as scatterer for fields emitted by the handset (i.e., the intended radiation [9]).

One way to avoid the undesired current is to use balun chokes or ferrite, as demonstrated in [9, 10]. However, this approach has some disadvantages:

- The chokes tend to be band-limited and thus need to be designed for the relevant frequency and changed accordingly.

- The chokes may be rather large and stiff in construction, so that natural handling of the device may be hindered during propagation measurements.

- Although the choke blocks the current, the cable is still there physically and hence fields emitted from the device may couple to the cable, which thereby can act as reflector. In addition, repetition of measurements may become difficult, since the influence of the flexible cable may be hard to repeat in practice. Repeated measurements may be needed for measurements involving users in order to estimate the influence of a user (see Section 7.3).

Apart from the above-mentioned problems, balun chokes can only to some extent be used in reflection measurements.

Another principal way of avoiding a conductive cable is to include a small transmitter acting as a reference source inside the DUT. Provided the extra source is well defined, the total transmitted power can then be measured in an anechoic room. However, these extra transmitters generally have narrow bandwidth and hence several may be needed to cover the relevant frequency band, thus making

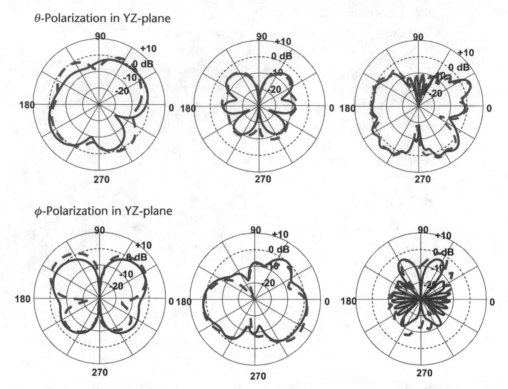

θ-Polarization in YZ-plane

φ-Polarization in YZ-plane

Figure 7.11   Simulated radiation pattern for the φ-polarization (top row) and the θ-polarization (bottom row) when a cable with a bend is connected. The dashed line is for the similar measurements. Similar to Figure 7.9, the different columns represent the frequency bands, 0.3–0.6, 0.6–1.2, and 1.8–3 GHz. (From: [11]. ©2001 from IEEE. Reprinted with permission.)

this approach less attractive. Especially for ultrawideband (UWB) antennas, the narrow bandwidth of the reference transmitter may be prohibitive. Furthermore, measurements of the phase are not possible with this setup (in principle).

An attractive alternative is to use analog transmission over an optical fiber [11]. Here, RF signals are converted into light using a laser diode. The light is guided in a flexible plastic fiber and finally converted into RF again using a photo diode. The loss in the conversion from RF into light and back again is large, but the output power is still sufficient for measurements in an anechoic room. The photo diode may be very small, and could for example fit into an SMA connector. Only a small button-cell battery is needed and therefore the transmitter part fits inside most small wireless devices. The bandwidth may be several gigahertz.

7.2.4   Total Radiated Power and Total Isotropic Sensitivity

For small devices like a mobile handset it is neither important for the user nor for the network where exactly a loss is introduced, such as in the antenna itself, due to a nonperfect matching, or due to load-pull of the power amplifier, and so forth. Furthermore, in testing of off-the-shelf products it may not even be possible to measure individual components in the transmitter chain.

The total radiated power (TRP) or equivalent isotropic radiated power (EIRP) is a measure of the radiation pattern including the matching losses and the output

Table 7.3   Measured TRP and TIS Values for Free Space and Including a Torso Phantom

		Handset				
		A	*B*	*C*	*E*	*F*
Free space	TRP	25.8	27.7	28.0	26.4	26.3
	TIS	−99.5	−98.2	−100.4	−99.0	−100.0
Phantom	TRP	23.2	24.2	25.9	23.6	23.5
	TIS	−96.4	−95.0	−98.5	−96.8	−96.4

Source: [42]. All values are in dBm.

power of the power amplifier on the device. Further, the measurements can be based on the peak burst power including various system aspects if the DUT is an off-the-shelf product. The matching losses are not only the loss due to standing wave ratio (SWR) stemming from imperfect impedance matching, but also the loss due to change in the power delivered by the amplifier due to a loading different from 50 Ω.

In the following the TRP is defined as the integral of the radiation gain pattern over all directions and both polarizations. Similarly, the total isotropic sensitivity (TIS) is the power necessary for the device receiver to operate at a threshold bit error rate (BER), assuming the power is distributed evenly over the sphere and between the polarizations.

From the radiated power measurement it is not possible to find the antenna efficiency but the radiation pattern itself is measured without any external cables and there will therefore not be any disturbance of the radiation pattern. Further, if limits are put on the maximal radiated power, this measure will include all parameters influencing the transmitted power and is the easiest way to obtain the peak or total transmitted power. By comparing the total radiated power to the input power for the antenna it is possible to obtain the total gain in dBi (dB relative to an isotropic radiation).

As an example of realistic TRP and TIS values, Table 7.3 lists the values measured on five commercial GSM handsets commonly used (circa 2000), and are described further in Section 7.4.2. Full spherical radiation patterns were measured for the handsets on the center GSM-1800 channel using the setup described in Section 7.5.1. The handsets were measured both in free space and next to a phantom.

7.3   User Influence

In mobile and wireless systems handheld phones experience significantly higher loss in the link budget than similar car-mounted phones. This higher loss is in the order of 10 dB or more, as shown in Section 7.1. The higher loss for handheld phones compared to car-mounted phones can be explained by two factors, namely, the different characteristics of the antenna radiation patterns and the human body being close to the antenna. A car-mounted antenna is typically a whip vertically mounted on the metallic surface outside the car, whereas the antenna on a handheld typically is a normal mode helix or PIFA mounted on the top/back of the handset. The main difference in the radiation pattern and polarization state between the car-mounted and the handheld antennas arises from the small bent ground plane in a handheld phone compared to the large flat ground plane on a car and the tilting of the phone when held in speaking position. The loss introduced by the presence of a user, the bodyloss, is taken as 3 dB in the GSM specifications, but from the measurements presented in Section 7.1 and elsewhere [2, 12, 13], a figure of 10 dB is commonly found.

The measured bodyloss covers only the loss due to the proximity of the human body by comparing the received power with the human to the received power for the phone with no user present, for the same location and tilt of the handheld phone. An interesting question is what causes the apparent loss in the link budget. The absorption by the human body is one contribution to the bodyloss, but simulations including a numerical head model has often shown that the absorption only accounts for 2 to 4 dB [14–17], indicating that other factors such as the antenna mistuning, change of radiation pattern, and change of polarization state due to the human proximity [15–17] have a more significant influence. To investigate the contribution to the bodyloss for each of the four factors: absorption, antenna mistuning, change of radiation pattern, and change of the polarization state for real live persons, a setup was built in a large anechoic room. By measuring the field strength on a sphere in the far-field of the transmitting phone held by a test person, the absorption can be found by simply integrating the radiated power and comparing it to the power delivered to the antenna. By using a directional coupler before the antenna on the phone both the forward and reflected power can be measured and the losses due to antenna mistuning can be found. The loss due to change in the direction of the radiation and polarization of the fields can be found from measurements of the radiation patterns for both polarizations of the phone held by a test person and a model, or measurement of the incoming multipath fields in a mobile environment. This will be described further in Section 7.4.

As discussed in Section 7.1 the bodyloss may vary significantly from one person to another, even for the same physical phone [18, 19]. The measurements in the anechoic room need therefore to involve a number of tests persons. In the campaign discussed below measurements with more than 40 persons were made in the anechoic room.

7.3.1   Experimental Setup

The anechoic room at Aalborg University is a rectangular room with dimensions $7 \times 7 \times 10$ m with absorbers on all sides. The suppression of reflections is better than 40 dB at 2 GHz. The room is equipped with a roll-over-azimuth-over-elevation pedestal with a fiberglass mast that can take 250 kg. The probe antenna (which is receiving) is normally located on a fixed mast some 5.5m from the pedestal. For measurements of the three-dimensional radiation patterns of a test user and the phone, using the existing setup the test person needs to be rotated in two dimensions. There are two major problems in that procedure: first, the test person might feel uncomfortable being rotated around two axes, and second, the position of the person relative to the phone is not allowed to change during the whole spherical measurement. One three-dimensional measurement takes about half an

Figure 7.12   The setup in the anechoic room for measurements of three-dimensional radiation patterns of a person using a handheld phone.

hour in the existing setup. To overcome this problem the probe antenna can be used to scan the sphere around the test person in elevation whereby the test person and phone only needs to rotate in azimuth. A construction for moving the two probe antennas (one for each polarization) was made (Figure 7.12).

The distance from the antenna on the phone to the probe antennas is 2.1m and the phone was placed in the center of rotation by moving the seated test person but before the measurements started, in both the X, Y, and Z directions. Placing the test person in the center of rotation was done using a laser pointer. The laser was also used to control if the test person was moving during the measurements. Two pictures were taken with the laser light on for every test person, one just before the measurement started and one when the measurement was completed (Figure 7.13). Furthermore, a video camera was recording each test person for the whole duration of the measurement for further study. Measuring one person takes about 22 minutes in this setup.

Due to the pedestal it is not possible to have a distance to the probe antennas larger than $L = 2.1$m, by using the far-field criterion this gives a test zone around the phone at 1.89 GHz of

Figure 7.13   A test person placed in the center of rotation and ready to be measured.

$$Test\ zone \simeq \sqrt{\frac{L\lambda}{2}} = 0.41\,m \qquad\qquad (7.1)$$

where λ is the wavelength. A test zone of 0.41m is sufficient to cover both the phone and head/hand of the test user, which is where nearly all current is located. To control the pedestal and movements of the probe antennas a two-axis Orbit controller was used. The Orbit controller was programmed to take one measurement for each 5 × 5 degrees on the sphere. To measure both polarizations and two antennas on the phone at the same time a logical circuit was made for handling the two PIN diode switches needed (Figure 7.14). The radio frequency measurement was made by a network analyzer measuring the S21 and by a power meter that was used to log the reflected power. The power reflected due to mistuning caused by the human user was measured eight times per second and logged by a PC together with the position of the switches. The velocity of the pedestal is 10° per second, leaving four measurements of the reflected power per 5°. In order for the power meter to settle, only the power reflected by one of the antennas was measured for each 5° movement and the last of these four measurements was used.

The reflected power is used for three purposes: to estimate the loss due to mistuning, in the calculation of the absorption, and to monitor if the test person is changing the way he or she holds the phone against the head. Just moving a finger on the back of the phone clearly affects the reflected power. If the test person is changing the way of holding the phone, the measurement is discarded.

The forward power was not measured during the measurements of the radiation and the changes in the forward power for different loadings of the amplifier were investigated later. The largest change in the forward power obtained by changing the position of the phone against the head and the hand was ±0.25 dB.

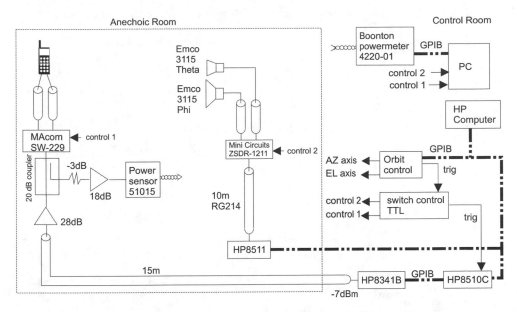

Figure 7.14   The setup used for the measurements of three-dimensional radiation patterns and reflected power for 44 test persons holding a handheld phone equipped with two antennas. (From: [39]. ©1999 IEEE. Reprinted with permission.)

The largest variation in the forward power obtained by using a sliding short was ±1.0 dB, again including the switch and the small length of cable from the directional coupler to the handset antennas (1.6 dB of loss). The sliding load correspond to moving on the circle where |Γ| = 0.72 in the Smith chart and is considered worst case.

The measured S-parameters are converted to the electric field $E_\theta(\theta,\phi)$ and $E_\phi(\theta,\phi)$ for each person for both antennas measured on the phone and the TRP was calculated by

$$P_{TRP} = \int_0^\pi \int_0^{2\pi} \left[\left| E_\theta\left(\theta,\phi\right) \right|^2 + \left| E_\phi\left(\theta,\phi\right) \right|^2 \right] \sin\left(\theta\right) d\theta\, d\phi \qquad (7.2)$$

as explained in Section 7.2.4.

7.3.2   Repeatability Test

The most critical issue in the measurements is if the test person changes the way of holding the phone during the 22 minute measurement. To investigate the repeatability of the measurements, six test users were asked to participate in two measurements. The test persons had a 5-minute break between the two measurements and they were asked to try to hold the phone in the same way for both measurements. The results of transmitted power relative to the best of the two antennas on the phone transmitting in free space, as well as the reflected powers are shown in Table 7.4.

As seen the transmitted power is within ±0.5 dB for the two measurements and the change in mean reflected power is also low, especially for the higher val-

Table 7.4   Transmitted Power in dB by Six Test Persons Measured Two Times

		Repeat test person					
		1	2	3	4	5	6
Helix	Transmitted power	−10.1	−9.7	−9.5	−9.3	−9.3	−8.8
		−10.9	−10.0	−9.5	−8.6	−10.2	−8.0
	Reflected power	−12.9	−10.6	−15.2	−14.5	−14.2	−9.2
		−13.1	−13.3	−15.2	−14.0	−12.2	−8.5
Patch	transmitted power	−3.4	−2.2	−1.7	−1.8	−2.2	−3.1
		−4.0	−1.5	−1.8	−1.8	−2.8	−3.5
	Reflected power	−5.7	−5.2	−5.7	−6.1	−6.1	−4.9
		−5.6	−5.0	−5.8	−6.9	−5.2	−4.8

Source: [39].

The transmitted power is relative to the power in free space, and the reflected power is relative to the input power.

ues, which are the most important ones as well as the most accurately measured. The worst change in reflected power corresponds to only 0.25-dB change in the power delivered to the antenna. The correlation coefficients between the radiation patterns are also calculated and they are approximately 0.96. From this test it is concluded that it is possible to use the setup within an overall uncertainty of 1 dB, and as the transmitted power by one person compared to another can change by 10 dB the setup is concluded to be sufficiently accurate.

7.3.3 Variation Due to Different Users

Altogether 44 persons were measured in the anechoic room setup, six of the test persons were measured twice to see how well the measurements can be repeated as well as how much small movements during a single measurement change the results. Furthermore six free-space measurements were made, three for the helix and three for the patch antenna. The three measurements are for different placements of the phone in the test zone and for different tilting angle of the phone. For eight of the measured persons an error occurred in the logging of the reflected power and these files were discarded, leaving 36 measured test persons of whom six were measured twice. In addition, six free-space measurements were made.

The reflected power for the antenna with a test person present and in free space is shown in Figure 7.15. The persons are sorted according to the average reflected power to ease the view and with the free-space measurements located as the last six measurements in the plot. The error bars show the minimum and maximum of the recorded values, which gives an idea of how stably the test persons hold the phone. For the low values of reflected power the change in reflected

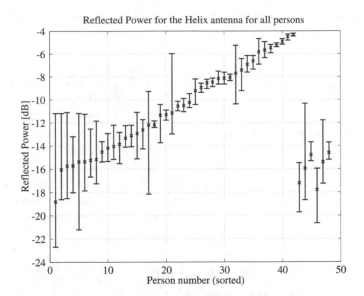

Figure 7.15 The reflected power for the helix antenna relative to the forward power at the antenna terminal. The error bars shows the maximum and minimum values during the 22-minute measurements. (From: [39]. ©1999 IEEE. Reprinted with permission.)

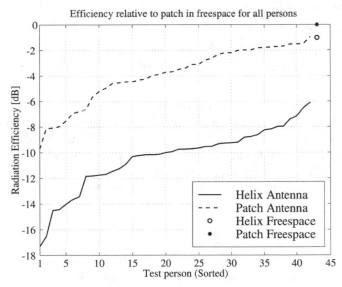

Figure 7.16 The radiation efficiency relative to the patch antenna in free space for all the test persons. (From: [20]. ©2000 IEEE. Reprinted with permission.)

power hardly changes the power delivered to the antenna. The loss due to mistuning of the antenna is always less than 2.0 dB for both antennas.

In Figure 7.16, the sorted radiation efficiencies for the test persons are shown together with the efficiencies for the free-space case. It is clear that the patch antenna results in a significantly lower absorption compared to the helix antennas. The average overall losses for the helix and the patch is 9.7 dB and 3.4 dB, respectively. Furthermore, there is a large correlation between the overall losses for the patch and the helix; the correlation coefficient is 0.80. For the free-space measurements the antenna with the lowest loss is the patch whereas the helix has 1 dB higher loss.

To summarize the above, there are four factors that contribute to the body-loss: absorption, antenna mistuning, change of radiation direction, and change of the polarization state. The major part of the bodyloss is found to be absorption, whereas the mistuning only accounts at most to 2 dB. The average absorption is 3.4 dB for the patch antenna and 9.7 dB for the normal mode helix antenna.

The bodyloss results presented in Section 7.1 involving 200 test persons in a mobile environment were obtained with the same patch and helix antenna. The bodyloss was found to be 3 dB for the patch and 10 dB for the helix antenna. This indicates that the change of radiation direction and change of the polarization state only plays a minor role in the bodyloss. So the question then is what causes the bodyloss to vary significantly between different users of the same handsets. The following is a list of possible reasons:

- Position of the hand on the handset;
- Distance between the head and the handset;
- Tilt angle of the handset:
- The shapes of the head and the hand;
- The size of the person;
- Variations in skin humidity;
- Many, possibly minor, parameters such as age, sex, amalgam teeth fillings, glasses, amount of hair, and so forth.

In [20], the first three items of the above list were investigated using measurements similar to the ones described above for test users, but using a phantom head and hand instead. Regarding the tilt angle, less than ±0.6 dB change in the radiation efficiency was found for tilt angles between 0 and 90°, while changes up to about 0.9 dB were found when the distance between the handset and the phantom head was changed from 0 to 35 mm. Thus both influences are small compared to the changes observed among the different users.

7.3.4 Influence of Hand Position

Early results on the influence of a handset user on the performance was presented in [17], based on simulations. In this work, the loss in the simple head plus hand phantom indicates that the loss due to the hand is very small compared to the loss in the head. In the simulations the hand covered all the handset except 50 mm at the top.

To analyze this topic further, four measurements were made with the phantom hand in different positions (Figure 7.17). The corresponding radiation efficiency results are shown in Figure 7.18. As can be seen from the figure, the absorption is practically the same for the hand holding the phone at the bottom and for the case of no hand. This is in good agreement with the simulations in [17], where it was found that only 4% of the absorption was in the hand. But if the hand is close to the top of the phone and thereby the antenna, this changes completely. If the hand

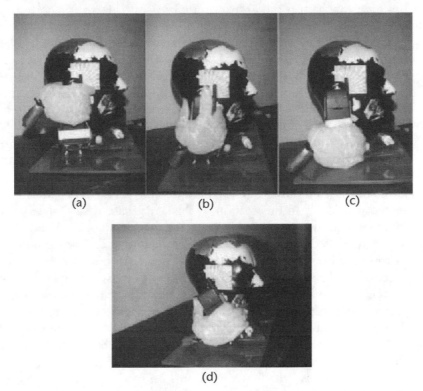

Figure 7.17   The four phantom hand positions. (a) The hand in top position, (b) the hand in near top position, (c) the hand in bottom position. The last position (d) is for the middle position. (From: [20]. ©2000 IEEE. Reprinted with permission.)

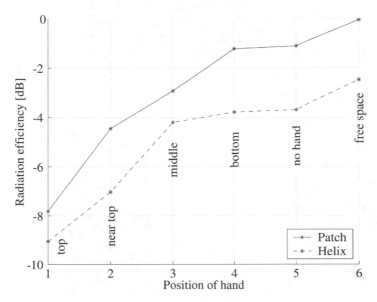

Figure 7.18   Loss due to the simple head plus hand phantom as a function of the position of the hand. (From: [20]. ©2000 IEEE. Reprinted with permission.)

covers most of the top the absorption increases to 8 dB for both antennas. This shows that the hand is the single most important part when considering the loss for this simple head plus hand phantom. Actually the simple head model only accounts for about 1 dB, whereas including the hand causes the losses to increase up to 8 dB, even without the hand touching the helix antenna. Note that the loss in the integrated patch does not exceed the loss for the helix even though the patch in one position is nearly totally covered. In addition, measurements of the loss for the helix when it is touched by either a finger or the ear show that the loss can increase dramatically, mainly because of absorption.

Summarizing the results for the measurements of the phone together with the simple head plus hand phantom, only the position of the hand can explain why the absorption from one person to another can vary by more than 10 dB, as demonstrated previously in this section. If this is the case it should be possible to link the absorption shown in Figure 7.16 with the position of the hand for the live persons measured in the anechoic room when they used the same phone.

During the live person measurements, each person was recorded on video, and at the back of the phone a thin tape with black and white squares of 10 × 10 mm was attached. Using the squares it is possible to locate the position of the hand very precisely just from the video. Each person is, of course, holding the phone in a unique way and in order to ease the investigation, the positions are quantified by the number of centimeters that are not covered by at least one finger, in the height of the phone (Figure 7.19).

Figure 7.20 shows the loss versus portion of the phone covered by the user's hand. As can be seen, the losses do correlate strongly with the position of the hand, the correlation coefficient between the loss and the position of the hand is 0.7 for the patch and 0.67 for the helix. It should be noted that only one person was holding the phone in the positions 3, 4, 8, and 12 cm, all other positions were used by at least three persons. The loss seems rather constant when the hand covers up to 10 cm, the critical part is when the finger enters the top 3 cm. Forty percent of the tested persons entered this region and more than two-thirds entered the top 4 cm.

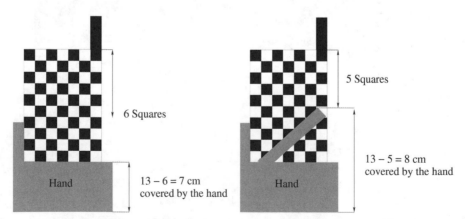

Figure 7.19   Examples of quantified holding positions for live persons. The grid is 10 × 10 mm. (From: [20]. ©2000 IEEE. Reprinted with permission.)

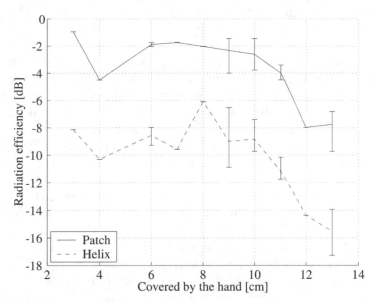

Figure 7.20   The loss due to live persons using the phone sorted according to the way they hold the phone. (From: [20]. ©2000 IEEE. Reprinted with permission.)

7.3.5   Summary

The loss due to the position of the hand on the phone is found to be the primary parameter whereas the tilt angle and distance relative to the head are minor parameters. For this rather large phone (length 13 cm) the hand causes high losses for one-sixth of the tested persons but for the more recent and much smaller phones the losses can be expected to be significantly higher for the majority of users, if nothing is done to prevent this. More than two-thirds of the tested persons entered the top 4 cm on the 13-cm-long phone and perhaps this is the percentage of users that will enter the top 2 cm for a 10-cm-long phone, the length of most "bar phones" on the market today.

It is also clear that the integrated patch antenna always has a lower loss than the normal mode helix antenna, when the phone is used in normal position. This is even true when the hand nearly covers the patch antenna totally. This was also the conclusion from the measurements in a real mobile environment, mentioned in the beginning of this section [21]. The loss due to the simple phantom consisting of a head and hand results in a significantly lower loss than is the case for real persons. It is therefore important to find a radiation phantom yielding results closer to those obtained with real persons. The measured results indicate that the hand is more important concerning radiation efficiency than the head, a new finding at the time. Due to smaller and smaller phones this makes it a challenge to design good antennas for future phones.

The above results obtained via measurements are supported by the work presented in [22]. Using FDTD simulations it was concluded that the presence of a hand/finger near the antenna has a large influence on the losses (e.g., 6–12 dB for a monopole antenna), and that a head phantom alone only gives about 4 dB of loss compared to free space. In addition, the simulation method was verified by comparing measurements and simulations of the handset and phantom/hand, showing an excellent match.

In a later work described in [23] possible reasons for the large variations in the bodyloss are investigated further. Using two GSM-900/1800 phones, about 100 mm long, measurements were made for the uplink (UL) in an anechoic room with 16 test persons and phantoms in various constellations. From a series of measurements where the test users were allowed to freely select the hand position on the phone, but not the side of head, large variations in body loss of up to 8 dB were found, similar to the previous reports mentioned above. By using a fixture, a series of measurements were also made where only the test user's head is near the phone, but not the hand. From this, a rather low average bodyloss of 5.4 and 2 dB was found for the 900- and 1,800-MHz bands, respectively, where the corresponding values for the phantom were 5.8 and 1.5 dB. Thus, the influence of the user's head is modeled quite well with standard deviation of 0.2 to 0.4 dB among the users. The variation in the bodyloss due to the hand was further investigated by measurements where the users held the phone in a predefined way (i.e., as much the same way as possible). The variations in body loss were then maximum 2.4 and 1.2 dB for the 900- and 1,800-MHz bands, respectively, with corresponding standard deviations 0.8 and 0.4 dB. Thus, it was concluded that the main contributor to the body loss variation is the difference in the hand position, verifying the results described above.

7.4 Mean Effective Gain

It is a characteristic of antennas in mobile or cellular systems that their performance depends on the environment. The environment consists of the user, which in some respects can be considered as part of the antenna, and the external environment. The multiplicity of waves usually incident on the user creates a Rayleigh fading channel, the distribution of which a single, one-port antenna cannot change. The average received power may be optimized using the average power distribution in the environment, including their polarization properties.

An expression for the complex signal at a matched antenna port was first given in [24] derived with the assumptions that:

1. The phase of the incoming electric field is independent of the arrival angle for both polarizations.
2. The phases of the incoming electric field in the two polarizations are independently distributed between 0 and 2π.

With these assumptions the voltage at the antenna port is

$$v(t,f) = \oint E(\Omega,f) \cdot A(\Omega,t,f) d\Omega \qquad (7.3)$$

where $E(\Omega;f)$ is the electric far-field pattern of the antenna, $\mathbf{A}(\Omega;t,f)$ is proportional to the electric field of the incident plane waves, and t indicates that the environment is changing with time (e.g., by movement of the user). In addition, the variable f indicates that the environment and the radiation patterns may depend on frequency. For simplicity a narrowband system is assumed, so that the frequency can be omitted below.

The average received power at the antenna is

$$P_{Rx} = \left\langle v(t) \overset{*}{v}(t) \right\rangle = \oint \left[G_\theta(\Omega) P_\theta P_\theta^n(\Omega) + G_\phi(\Omega) P_\phi P_\phi^n(\Omega) \right] d\Omega \qquad (7.4)$$

where $\left\langle v(t)\overset{*}{v}(t) \right\rangle$ denotes averaging in time, and

$$P_\psi = \oint \left| A_\psi(\Omega) \right|^2 d\Omega \, ; \qquad P_\psi^n(\Omega) = \frac{\left| A_\psi(\Omega) \right|^2}{P_\psi} \qquad (7.5)$$

is the total power and normalized power distribution, respectively, where ψ stands for either θ or ϕ. $A_\psi(\Omega)$ is the ψ-polarization component of $A(\Omega)$, and $G_\psi(\Omega) = |A_\psi(\Omega)|^2$. The antenna gains are normalized as usual,

$$\oint \left[G_\theta(\Omega) + G_\phi(\Omega) \right] d\Omega = 4\pi \qquad (7.6)$$

By also normalizing the powers Taga arrived in [25] at the very useful definition of the mean effective gain (MEG)

$$\Gamma = \oint \left[\left(\frac{\chi}{1+\chi} \right) G_\theta(\Omega) P_\theta^n(\Omega) + \left(\frac{1}{1+\chi} \right) G_\phi(\Omega) P_\phi^n(\Omega) \right] d\Omega \qquad (7.7)$$

where the cross-polar discrimination (XPD) χ is defined as

$$\chi = \frac{P_\theta}{P_\phi} \qquad (7.8)$$

The MEG is a normalized measure of the received power, equal to one-half for isotropic antennas where $G_\theta = G_\phi = 1/2$. The optimal handset antenna will be the one that maximizes the MEG for the relevant environmental scenarios. Antenna designers are used to stringent requirements in free-space environments with a single-wave incident, maximizing gain, minimizing sidelobes, and so forth. It is therefore a sobering observation that designing handset antennas for a totally random environment, where $\chi = 1$, $P_\theta^n = P_\phi^n = (4\pi)^{-1}$ is in principle easy, since in this case the MEG is

$$\Gamma_{random} = \oint \left[\frac{1}{2} G_\theta (\Omega) \frac{1}{4\pi} + \frac{1}{2} G_\phi (\Omega) \frac{1}{4\pi} \right] d\Omega = \frac{1}{2} \qquad (7.9)$$

and thus independent of the radiation patterns. These assumptions are not exactly valid in practice as will be seen, but they do indicate that the radiation pattern is not the most important parameter to optimize. More realistic models of the incoming power are presented in the following section.

7.4.1   Power Distribution Models

As it will be apparent from the results presented in later sections, the MEG in (7.7) is highly dependent on the models $P_\theta^n(\Omega)$ and $P_\phi^n(\Omega)$ of the power distribution versus angle in the two polarizations, and in addition the associated XPD, X. Therefore, the models applied should be chosen carefully and should be based on observations of how the power is distributed versus the azimuth and elevation angles in typical situations. This section will first describe how such a model was derived from measurements, followed by a summary of a few models that have been discussed in the literature.

An important scenario for mobile phones is where the base station (BS) is located at an elevated position in an urban environment and the mobile station (MS) is inside a building. In order to derive a model for this scenario a series of measurements were conducted with MS locations on all floors of a four-story building in the city of Aalborg, Denmark. Five different BS locations were used, placed in a ring around the building, with heights and distances to the measurement building as summarized in Table 7.5. The BS locations were chosen to mimic a typical cellular network; the BSs were equipped with vertically polarized 60° sector antennas, with a vertical beamwidth of about 5°.

A correlation-based channel sounder was used to measure the complex channel impulse response in a 20-MHz band around 1,890 MHz, although in the work described below only the wideband power is used. The sounder had an instantaneous and overall dynamic range of about 45 and 80 dB, respectively.

With the purpose of measuring the received power distribution versus azimuth and elevation angle, a dual-polarized narrowbeam horn antenna was used

Table 7.5   Overview of the BS Sites, Heights over the Ground, and Approximate Distances to the Rx Site

Site	No.	Distance	Height
Sygehus Nord	1	450 m	40 m
Rolighedsvej	2	1350 m	35 m
Nordkraft	3	1400 m	42 m
Mølleparken	4	1450 m	50 m
Havnen	5	1400 m	2 m

Source: [27].

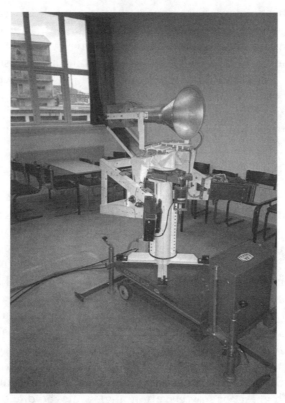

Figure 7.21   Horn antenna mounted on the device turning the main beam into different combinations of azimuth and elevation angles.

at the Rx end of the sounder. The horn antenna was mounted on a positioning device controlled by step motors. The device allows steering the horn antenna direction into a desired combination of azimuth and elevation angles, as controlled by an acquisition program collecting the measurements. For each position of the Rx a series of impulse response measurements was obtained with the antenna pointing in 72×15 directions (i.e., all combinations of the azimuth angles $\phi\in\{0°, 5°, 10°, ..., 355°\}$ and the elevation angles $\theta\in\{40°, 45°, ..., 110°\}$, where the angles are given in usual spherical coordinates). The impulse responses for both the θ– and the ϕ– polarizations were measured simultaneously using two channels of the sounder. Each series of such measurements characterizes in 3-D the actually received signals as seen by any antenna located at the same position. For these measurements it is assumed that no signals arrive from directions outside the measured set of directions and furthermore that the channel is essentially constant during the measurement duration of about 2 minutes. The mounted horn antenna is shown in Figure 7.21.

The radiation patterns of the horn antenna are shown in Figure 7.22. Compared to the beamwidth of the horn antenna the increment of only 5° in both the azimuth and elevation angle is rather small. Therefore the measurements in the different directions will be dependent. However, the intended application of the measurements and the model is to compute MEG for mobiles with comparably small antennas, and hence generally large beamwidths, and therefore the angular resolution is sufficient. This conclusion is supported by the work described in [26] where the effect on MEG of deconvolving the horn antenna pattern was investigated.

The rooms used on all the five floors have windows in the same direction, except on the first floor, which has windows in the opposite direction. From the measurements it is a general observation that most of the power comes from the direction of the windows, so that the direction of the maximum power falls within an azimuth angle relative to the windows of ±60°, depending on the Rx and BS locations. For this reason the MEG of a handset may vary depending on the orientation with respect to the windows in the room. In order to preserve this important aspect in the model, all the measurements are aligned in azimuth, so that the maximum power is for an angle of 0°. In total measurements for 60 combinations of BS and MS locations were obtained, of which three were discarded due to insufficient SNR. Figure 7.23 shows the mean power distribution obtained after aligning and normalizing all measurements. The power is shown as planar plots where different shades of gray represent the power level, the radial axis represents the elevation angle with 40° at the center and 110° at the edge, and each circle corresponds to an azimuth scan 0° to 355°. The XPD for the mean model was computed according to (7.8) as 5.5 dB.

In addition, Figure 7.24 shows the mean power distributions for the θ– polarization versus azimuth and elevation angles, including bars indicating the standard deviation of the individual measurements. It is observed that the standard deviation is lowest in azimuth around the direction with maximum power, indicating that the measurements are more similar in this direction. The standard deviation is highest for low elevation angles, maybe due to ground reflections in some cases. Similar plots for the ϕ–polarization show a more even distribution of standard deviations versus direction, possibly because power in this polarization is only received via cross-coupling in the environment.

The average power distributions shown in Figure 7.24 can be used directly for computing MEG from (7.7). It is convenient, however, to derive parameterized models to facilitate the model description, and in addition this may provide deeper insight into the propagation mechanisms in the mobile channel. The parameterized models may be based on theoretical considerations or be derived from more or less empirical observations. Below a brief overview of existing models is given, with literature references providing more detailed information.

Although the description above has focused on the downlink (DL) direction (i.e., transmission from a BS) and reception by an MS inside the building, the radio channel is reciprocal, and the model may be applied for the UL as well. In this way the interpretation of $P_\psi{}^n(\Omega)$ depends on the link direction. For the DL, $P_\psi{}^n(\Omega)$ is the average normalized power incident on the handset from the direction Ω in the ψ–polarization. For the UL, $P_\psi{}^n(\Omega)$ is the normalized power received on average by the BS stemming from the MS transmitting in the Ω direction and in the ψ–polarization.

7.4.1.1 AAU

A model based on numerous outdoor to indoor measurements in the city of Aalborg, Denmark [27], as described above. This model is nonuniform versus both azimuth and elevation angle, and has an XPD of 5.5 dB.

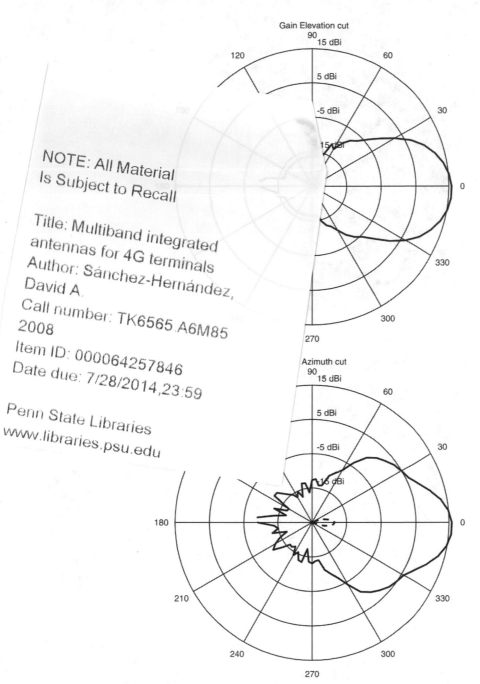

Figure 7.22 Gain patterns of the dual polarized horn antenna. The gain is 15.6 dBi, the 3-dB beamwidth is 26° for the azimuth cut and 31° for the elevation cut, the front-to-back ratio is 25 dB and the XPD is better than 40 dB. (From: [27]. ©2000 IEEE. Reprinted with permission.)

7.4.1.2 HUT

A model based on numerous outdoor to indoor measurements in the city of Helsinki, Finland [28]. In this model the variation versus azimuth angle is assumed uniform and nonuniform versus elevation angle. It has an XPD of 10.7 dB.

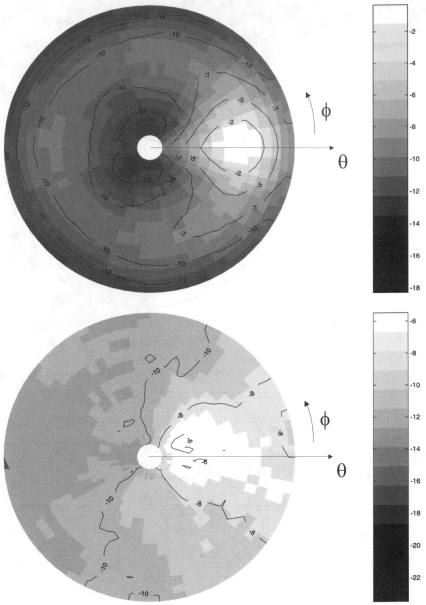

Figure 7.23 Mean of all spherical power spectrum measurements in the θ polarization (top) and
φ polarization (bottom), both normalized according to (7.5). The XPD is 5.5 dB. All
values are in decibels. (From: [27]. ©2002 IEEE. Reprinted with permission.)

7.4.1.3 Rect0

The rectangular model proposed in [29] has uniform weighting inside the window
defined by $45° \leq \theta \leq 135°$ for all ϕ, and zero weighting outside this window, where
θ is the elevation angle measured from the vertical axis and ϕ is the azimuth angle.
The XPD is 0 dB for this model.

7.4.1.4 Rect6

Similar to Rect0, but with an XPD of 6 dB.

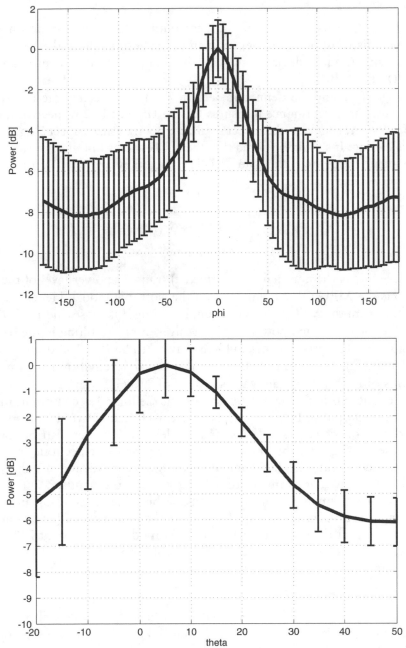

Figure 7.24   Measured mean power distribution in θ–polarization versus azimuth angle (top) and elevation angle (bottom), all normalized to a peak power of 0 dB. The standard deviation, in decibels, of all measured distributions are shown by error bars. (From: [27]. ©2002 IEEE. Reprinted with permission.)

7.4.1.5   Iso

The hypothetical isotropic model implies equal weighting of power versus direction in both polarizations and with an XPD of 0 dB. This model results in MEG values equivalent to the TRP and TIS, for the UL and DL, respectively.

7.4.1.6 Taga

This model was proposed by [25] for outdoor propagation and has a Gaussian power distribution in the elevation angle for both the θ– and ϕ–polarization (with different standard deviations), while it is uniform in the azimuth angle. It has an XPD of 5.1 dB.

The TRP and TIS defined in Section 7.2.4 have been suggested as initial handset antenna performance measures for the UL and DL, respectively, but the TRP/TIS does not include the directional and polarization aspects, and hence may be misleading. It should be noted that the MEG values obtained in the isotropic environment differ from the true TRP/TIS since they are based on the antenna gain patterns. The TRP is defined as

$$P_{TRP} = \frac{P_{Tx}}{4\pi} \oint_S \left[G_\theta(\Omega) + G_\phi(\Omega) \right] d\Omega = 2P_{Tx}\,\Gamma_{Iso} \qquad (7.10)$$

where P_{Tx} is the nominal (or conducted) transmit power level of the handset and Γ_{Iso} is the MEG value obtained with the isotropic model. Hence, the TRP can be found from the MEG value via a simple scaling. Likewise, the TIS is $P_{TIS}=P_c/(2\Gamma_{Iso})$ where the conducted power (at the receiver input) resulting in the receiver operating with a bit error rate of 2.44% is defined to be $P_c=-102$ dBm. In order to ease comparison, the results below for the isotropic environment are presented in terms of Γ_{Iso}, (i.e., as gain values instead of TRP/TIS values).

The Rect0 model was proposed by the Cellular Telecommunications & Internet Association (CTIA) in [29] as a "Near-Horizon Partial Isotropic Sensitivity" and may be viewed as a very simple model of the power distribution in the environment. Although this model does not appear to be accurate in many cases, it does incorporate the fact that in many mobile environments the power is not likely to arrive for example from directly above the handset. On the other hand, dense urban environments with high-rise buildings may result in power coming from reflections or refractions high above the user handset [25]. In addition it can be noted that the nonzero part of the Rect model covers about 71% of the sphere area. The Rect6 model is a simple attempt to add some typical polarization aspects into the Rect0 model.

7.4.2 Single Antenna

Using the models of the directional power distribution described in the previous section, the MEG can be computed from (7.7) if the spherical radiation pattern of a handset is known. The radiation pattern may be simulated in software using, for example, the finite difference time domain (FDTD) technique, or it could be measured from a final product or prototype. For example, as described in Section 7.3, series of spherical radiation pattern measurements were made with a mockup handset equipped with both a normal mode helix and an internal patch antenna. These measurements were performed including 42 different users holding the handset to the ear.

Since the measured radiation patterns to some degree are directive and at least some of the models described in Section 7.4.1 are also nonuniform it follows that the MEG depends on the orientation of the device in the model environment. Figure 7.25 illustrates the variation in MEG when the user is turning 360° in azimuth. The polar plot shows the variation in MEG for the average measured power distribution discussed in Section 7.4.1 (see Figure 7.23), the statistical model derived from the same measurements, denoted the AAU model in Section 7.4.1. Finally, results for the Taga model are also included.

The curves for the average measured power distribution show variations in MEG of about 3.5 and 4.5 dB for the patch and helical antennas, respectively, and the curve for the AAU model follows the curve for the measured data quite closely. The Taga model, on the other hand, is uniform in azimuth and hence yields a constant MEG when the handset and user are turning. Clearly, the Taga model intended for outdoor scenarios does not incorporate this important MEG variation found in the indoor environment.

Below, a comprehensive investigation into four GSM handsets is described where the variation of the MEG is analyzed for different orientations, both versus azimuth and elevation angle, and while using the different models described in Section 7.4.1. The four GSM handsets are commercially available and the spherical radiation patterns for both UL and DL have been measured as described later in Section 7.5. Handsets A and B are large handsets with external and internal antennas, respectively. Handsets C and F are small handsets with internal and external antennas, respectively. Here "small" handsets are about 10 × 4.5 cm, and "large" handsets are about 13 × 4.5 cm. Handset F was also measured with a substitute antenna (a retractable dipole); these measurements are labeled handset E. Although the same labels A, B, C, have been used, these handsets are not identical to the ones shown in Figure 7.6.

All the handsets were measured on the GSM-1800 channels 512, 698, and 885, corresponding to about 1,805, 1,842, and 1,880 MHz for the DL, respectively, and about 1,710, 1,747, and 1,785 MHz for the UL. These channels are the center and two edge channels of the GSM-1800 frequency band. The spherical radiation patterns were sampled using increments of 10° in both the azimuth angle ϕ and the elevation angle θ. The handsets were measured both in free space and next to a phantom head (Schmid & Partner v3.6), which was filled with a tissue simulating liquid [30]. For the free-space measurements the handsets were oriented along the z-axis of the coordinate system with the display pointing towards the negative y-axis. When the phantom was included, the handset was mounted on the left side of the phantom head at an angle of 45° from the z-axis, still with the display side facing the negative y-axis (see Figure 7.26). An example of a measured power gain pattern is shown in Figure 7.27.

As mentioned above, for real handsets in actual use both the radiation pattern and the spherical power distribution are directional and the MEG will vary depending on the orientation of the handset with respect to the environment. Therefore, it is necessary to consider more than one MEG value and, for example, compute a mean value. In addition, the variation in MEG for an environment is of interest; for example, the MEG might be unacceptably low in some cases although the mean is acceptable.

In order to investigate the variation in MEG, the measured radiation patterns have been rotated first with an angle of λ about the y-axis, corresponding to the phantom either bending forward or backward, and afterwards with an angle μ about the z-axis, corresponding to the phantom turning around in azi-

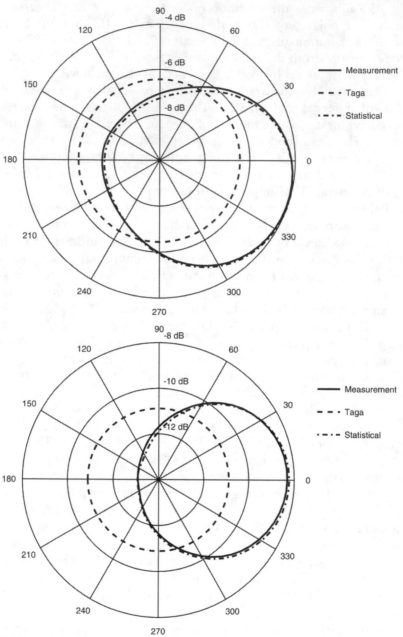

Figure 7.25 Calculated MEG curves using the measured spherical radiation patterns for a single typical test person with the mockup mobile phone and using either the average of all measured spherical power spectrum measurements (solid lines), the Taga model (dashed lines), or the AAU model. The top set of curves is for the patch antenna of the handset and the bottom set of curves is for the helical antenna. (From: [27]. ©2000 IEEE. Reprinted with permission.)

muth. For each desired point of the rotated pattern the coordinates for the corresponding point in the actually measured pattern are found and the measured θ– and φ–polarization components are mapped according to the rotation and the coordinates. As samples are needed from directions not in the original sam-

Figure 7.26   A handset mounted on the phantom.

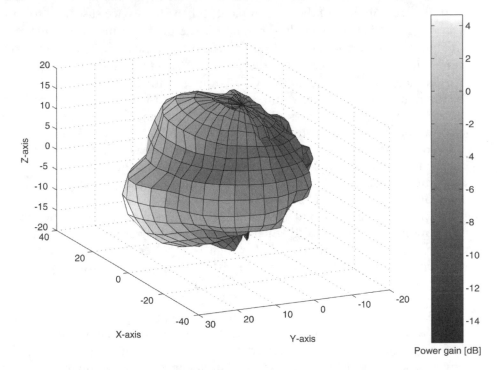

Figure 7.27   Measured power gain pattern for handset C including the phantom and measured on the center channel in the DL.

pling grid, spline interpolation has been used to obtain the rotated radiation patterns. All combinations of $\mu\in\{0°, 15°, 30°, ..., 345°\}$ and $\lambda\in\{0°, 15°, .., 60°, 300°, 315°, .., 345°\}$ have been used and for each combination of λ and μ the MEG was computed. Note that for the phantom measurements the described postprocessing rotation procedure corresponds to a simultaneous rotation of both the handset and the phantom.

The repeatability of the measurements has been investigated in order to determine the level of uncertainty in the MEG computations. For two different handsets (handset A and C) the maximum difference in the MEG computed from nine repeated free-space measurements was 0.2 to 0.4 dB, depending on the handset, link direction, and environment model. The values were computed as an average for the different orientations of the handsets. For three repeated measurements including the phantom the maximum difference was found to be 0.1 to 0.3 dB.

The accuracy of the measured values has been verified by the work in [22], where measurements made in the anechoic room on a monopole and a patch antenna were compared to FDTD simulations of the same antennas. A good agreement was found between the simulated and measured radiation patterns both having the same overall shape. The absolute value in each direction and polarization was usually within a few decibels. This is acceptable because the deviations are averaged in the MEG computation and because the largest deviations are where the absolute power level is low.

7.4.2.1   MEG for Different Environments

Figure 7.28 shows the MEG values obtained with the handsets in combination with different models of the environment, for the free-space case, and for both the UL and DL. The corresponding results for the phantom measurements are given in Figure 7.29. Furthermore, all results presented in this section are based on measurements made at the center of the GSM-1800 band, channel 698.

As mentioned above, the MEG is computed for many different orientations of the handsets in order to evaluate both a mean MEG and to estimate the variation in the MEG as function of orientation. In the figures the minimum and maximum of the computed MEG values are shown as the endpoints of a vertical line, one line for each handset. Also shown on each line is the mean value (shown with ×) and a MEG value for a specific orientation, marked with '□' (commented below). The results are presented in groups, one for each of the environments defined in Section 7.4.1.

Comparing the TRP and TIS results with those obtained using the rectangular window model (XPD of 0 dB) it is noticed that the results are very similar. The mean values are roughly identical, which is expected since the rectangular window covers about 71% of the sphere surface area. Hence, most of the power will be included, and as the XPD is zero no polarization weighting is used. Therefore the results will essentially be the TRP/TIS results.

Because the measured radiation pattern is rotated up to 60° in elevation angle some variation in the MEG values is observed for the rectangular window model, but only small changes are noticed compared to the changes seen with the two environment models derived from measurements. The rectangular window model with an XPD of 6 dB causes more changes, but the results are still far from those obtained with the AAU and HUT models.

Although the results obtained with the AAU model and the HUT model have some similarities, it is also clear that there are significant differences in some cases. For example for handset E in the free-space case the two models result in a MEG variation of about 2.4 and 5.5 dB in the DL direction for the AAU and HUT models, respectively, and about 3.1 and 7.7 dB for the UL direction.

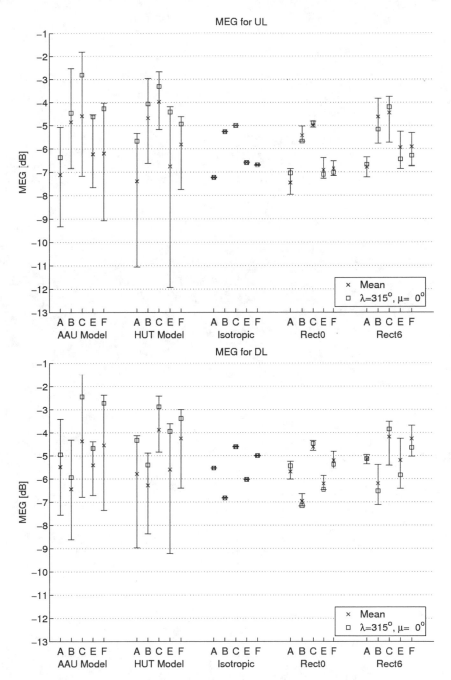

Figure 7.28   MEG for free-space conditions for both the UL (top) and the DL (bottom). The vertical line endpoints indicate minimum and maximum values. (From: [42]. ©2006 IEEE. Reprinted with permission.)

The large difference for the two models illustrates that both the power distribution versus angle and the XPD are important. The fact that before rotation handset E has a null near $\theta=90°$, where the HUT model has most of the power concentrated, combined with a high XPD of both handset E in free space and the HUT model yields MEG values that are highly dependent on the orientation.

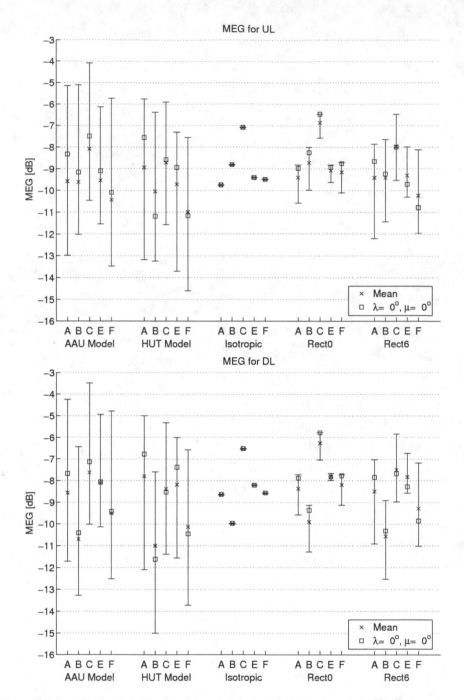

Figure 7.29   MEG for handset including phantom for both the UL (top) and the DL (bottom). The vertical line endpoints indicate minimum and maximum values. (From: [42]. ©2006 IEEE. Reprinted with permission.)

Table 7.6 shows the XPD for all the measurements, where it is noticed that the XPD may differ for the DL and UL. Hence, the frequency difference between the DL and UL must play a role.

When comparing Figure 7.28 and Figure 7.29, it appears that the measurements including the phantom generally have a larger difference between minimum

Table 7.6   XPD in dB for the Measured Radiation Patterns

	Free Space		Phantom	
Handset	DL	UL	DL	UL
A	3.7	5.2	0.3	1.3
B	8.2	8.7	-2.6	-2.5
C	3.8	4.4	-5.0	-3.8
E	13.7	11.8	1.6	-2.4
F	8.9	9.2	-5.9	-5.6

Note that the rotation of the patterns changes the XPD.

Source: [42].

Table 7.7   Difference in Mean MEG Values with the Isotropic Case as Reference

Environment	Handset, DL					Handset, UL				
	A	B	C	E	F	A	B	C	E	F
AAU	0.1	-0.7	-1.1	0.1	-0.9	0.2	-0.8	-1.0	-0.1	-0.9
HUT	0.8	-1.0	-1.9	0.0	-1.6	0.8	-1.2	-1.6	-0.3	-1.5
Isotropic	0.0	0.0	0.0	0.0	0.0	0.0	0.0	0.0	0.0	0.0
Rect0	0.3	0.1	0.3	0.4	0.4	0.3	0.1	0.2	0.3	0.3
Rect6	0.2	-0.6	-1.0	0.4	-0.7	0.3	-0.6	-0.9	0.1	-0.7

All values are in decibels for the phantom case.

Source: [42].

and maximum MEG than free-space measurements. This is due to the fact that the phantom blocks some of the power and effectively makes the radiation patterns more directive than the free-space patterns. This causes additional changes in the MEG when the handset is rotated and the environment model is directive.

Table 7.7 shows the differences in the mean MEG values obtained with various environment models compared to the TRP/TIS (i.e., isotropic environment). Measurements in Table 7.7 include a phantom. A similar table for the free-space situation shows that all differences follow within the range −0.3 to 1.0 dB.

The mean values are also quite small with the phantom in case of the rectangular window model with an XPD of 0 dB, where all differences are smaller than or equal to 0.4 dB. However, for the other models larger differences are found. In particular the HUT model results in differences from −1.9 up to 0.8 dB.

It is important to realize that even if the mean values are identical for two different models of the environment, the MEG values obtained with the two models need not be identical for a specific rotation of the radiation pattern. For the free space an example is the rotation of the measured radiation pattern with $\lambda=315°$ and $\mu=0°$, corresponding to a tilt angle of 45° in typical talk position. The MEG values obtained with these rotations are shown on the vertical lines in Figure 7.28

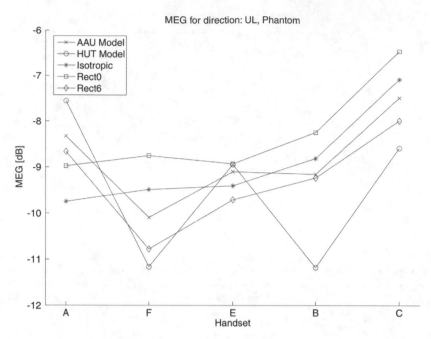

Figure 7.30   The MEG for a specific orientation ($\lambda=0°$ and $\mu=0°$) sorted for increasing TRP. All values are for the phantom case. (From: [42]. ©2006 IEEE. Reprinted with permission.)

with a □. It is clearly not possible to predict the MEG values shown with the □-marks from the mean values. The same is also true for the phantom measurements (Figure 7.29). For the phantom measurements $\lambda=0°$ and $\mu=0°$ is used since the handset is already mounted at an angle of 45° on the phantom. The MEG values obtained with the different models are depicted in Figure 7.30, where the MEG is shown sorted for increasing TRP. Each line in the plot represents a model of the environment and the values are all obtained with the same orientation of the handset/phantom as used above. The MEG for the AAU, HUT, and Rect6 models are not monotonically increasing, showing that the MEG values for these models cannot be predicted from the TRP values. As expected, the results for the Rect0 model are roughly similar to the TRP values, except for a constant offset.

7.4.3   Multiple Antennas

It is common to have receive space diversity at base stations, while only a few systems allow multiple antennas on the handset for reasons of complexity. There are some basic differences, since for the base station the signals are usually arriving over a limited range of angles, and it is therefore necessary to separate the antennas by many wavelengths. The two power levels will be equal on average. For the handsets the signals will be arriving over a large range of angles, making it fairly easy to obtain uncorrelated signals, while the two antennas may be of a different nature leading to different mean values of the power, and usually the orientation of the handheld phone with respect to the environment is changing with time. More importantly, the presence of a user will affect both the correlation coefficient and the MEGs (branch powers).

The diversity gain depends on the combining method and the probability level chosen. Figure 7.31 shows how the diversity gain for both selection combining and maximal ratio combining at the 99.5% level depends on the correlation coefficient between the branches and the power branch difference in decibels. Roughly speaking, the correlation coefficients should be below 0.5 and the powers within 5 dB to realize the maximum combining gain within a few decibels for selection combining (SC).

By using the lookup table in Figure 7.31, the diversity gain can be obtained for any correlation and branch power difference. The depicted selection combining results are obtained from close form expressions in [31], whereas the maximal ratio combining results are obtained by simulations of two Rayleigh-distributed signals with a certain correlation and branch power difference. For each simulation one element in the lookup table is obtained.

The envelope correlation thus needs to be low in order to obtain a worthwhile diversity gain, but it is difficult to give a closed form expression for the envelope correlation. For Rayleigh-fading signals the envelope correlation is closely related to the power correlation coefficients [31]

$$\rho_{env} \simeq \frac{\left| R_{xy} \right|^2}{\sigma_x^2 \, \sigma_y^2} \tag{7.11}$$

where R_{xy} is the cross-covariance, σ_x^2 and σ_y^2 are the variances (or average power) of the complex envelopes of antenna X and antenna Y, respectively. By derivations similar to those leading to (7.7), the correlation can be expressed as [32]

$$\rho = \frac{\oint \left[\chi E_{\theta X}(\Omega) E_{\theta Y}^*(\Omega) P_\theta^n(\Omega) + E_{\phi X}(\Omega) E_{\phi Y}^*(\Omega) P_\phi^n(\Omega) \right] d\Omega}{\sqrt{\oint \left[\chi G_{\theta X}(\Omega) P_\theta^n(\Omega) + G_{\phi X}(\Omega) P_\phi^n(\Omega) \right] d\Omega \oint \left[\chi G_{\theta Y}(\Omega) P_\theta^n(\Omega) + G_{\phi Y}(\Omega) P_\phi^n(\Omega) \right] d\Omega}} \tag{7.12}$$

where $E_\psi\chi$ and $G_\psi\chi$ are the radiation pattern and gain pattern for antenna X in the ψ-polarization, and similarly for antenna Y. The XPD χ is defined in (7.8).

Figure 7.32 shows the correlation coefficient between different antennas on one handset calculated from (7.12) and turning in different environments with and without the presence of a human head-model. The three antenna configurations investigated are two dipoles on top of the handset, a dual polarized patch antenna on the back of the handset and one dipole on top and a patch on the back of the handset (Figure 7.33). The three diversity configurations are selected from the criterion that both omnidirectional-like and more directional antennas, as well as a combination, have to be covered to investigate the correlation behavior.

The complex far-field patterns used in (7.12) are obtained from FDTD simulations. Four FDTD simulations were performed for each configuration. Two simulations are done in free-space, one for antenna X and one for antenna Y, and two simulations in the presence of the human head. In each simulation one antenna is excited while the other antenna is open-circuited. The dual polarized patch antenna has only one antenna port and therefore no open circuit exists. The polarization of the dual polarized patch antenna is selected by switching PIN diodes mounted on two consecutive sides of the antenna element and the ground on and off. This antenna has several advantages; it is compact, no switch is situ-

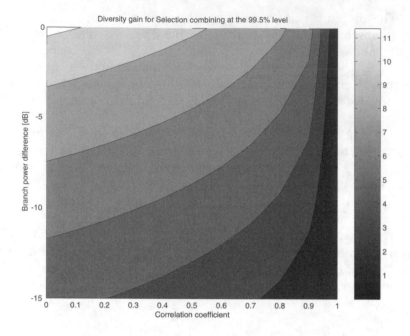

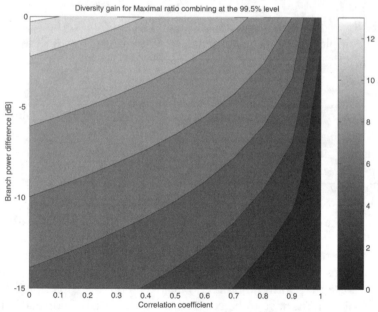

Figure 7.31   Diversity gain for selection combining (top) and maximal ratio combining (bottom) at the 99.5% level as a function of correlation between the branches and branch power difference. (From: [47]. ©1999 Union Radio-Scientifique Internationale [URSI]. Reprinted with permission.)

ated in the receiving or transmitting chain and the antenna impedance does not change depending on which polarization is active. This antenna configuration is of course only feasible for switching combining.

In general it is observed that the correlation in free space is usually very low, less than 0.4, and lower for dissimilar antennas. When the user head is present the

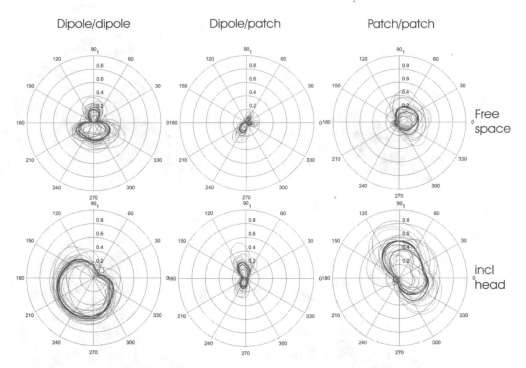

Figure 7.32   Correlation coefficients for three antenna configurations, on the top without a user and the bottom with a model of a user's head. The three thick curves in each sub-plot display the average correlation in a specific environment whereas the thin curves represent the correlation for one specific location within one environment. (From: [48]. ©1997 IEEE. Reprinted with permission.)

correlation rises considerably for the similar antennas, up to 0.6. The reason is that the shadowing of the head leads to a spectrum of waves incident from a more narrow range of angles, and thus an increased correlation. Should the signals come from only one direction, they would be highly correlated. The dissimilar antennas are still uncorrelated, indicating that the different radiation patterns are effective. Note also that the correlation in the different environments is quite similar although the environments are very different (an office environment, a railway station, and a large shopping mall).

Having the correlation between the two antennas (from calculated or measured complex far-field patterns and the distribution of incoming multipath fields) and the ratio of the mean received signal power by the antennas, the MEG values, and the diversity gain can be found for a chosen combining method and for the given environment. The diversity gain for a combining method can be obtained by either closed-form expressions, link simulations, or measurements.

Closed form expressions exist only for a few combining methods and often special requirements related to specific systems of interest leave only simulations or measurements of the diversity gain as an option. What is needed is a lookup table of correlation, difference in branch power and diversity gain as the one shown in Figure 7.31 for the combining method of interest. For example, using link simulations of the DECT system with a given speed and combining algorithm, this lookup table can be obtained by one simulation for each pair of correlation and branch power difference of interest [33]. Similarly, measurements can

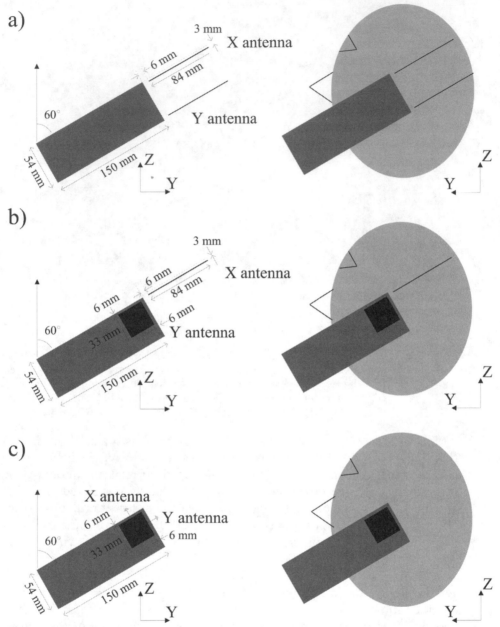

Figure 7.33   (a–c) The three antenna configurations examined. The handset has the same dimensions but is either equipped with two dipole antennas on top or one dipole on top
and a patch on the back or a dual patch on the back. All configurations are examined with and without a model of user's head. (From: [48]. ©1997 IEEE. Reprinted
with permission.)

be conducted to obtain the lookup table, but this requires the hardware combiner
and two radio channel emulators.

When the lookup table is obtained the diversity gain is found by a simple
lookup using the correlation and branch power difference as the entry. As an
example the dipole-patch configuration shown in Figure 7.33 is evaluated. First,
the handset is either built and 3-D complex far-field patterns are measured or sim-

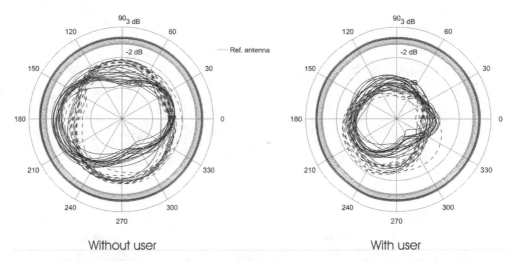

Figure 7.34   MEG values for both the dipole and patch antenna shown in Figure 7.33(b) when the handheld is turned 360°. The solid lines represent the patch antenna and the dashed lines represent the dipole antenna. The difference between the solid and dashed lines for each direction gives the branch power ratio. The handheld and the BS are located indoor in the same building. (From: [47]. ©1999 Union Radio-Scientifique Internationale [URSI]. Reprinted with permission.)

ulations are made to obtain the 3-D complex far-field patterns. Next, the environment is either measured to obtain the distribution of 3-D incoming multipath field in both polarizations or a model describing the distribution of incoming multipath field is used, as in Section 7.4.1. By using the MEG expression in (7.7), the received power gain for both antennas can be calculated while the user is turning in the environment. In Figure 7.34 the solid lines represent the MEG of the patch antenna and the dashed lines represent the MEG of the dipole antenna in the measured environments. The distance between the solid and dashed lines shows the branch power difference for each direction.

To obtain the correlation while turning in each environment, the 3-D complex far-field patterns and the 3-D incoming multipath field for each environment are again used, now in (7.12). The results are shown in the center column of Figure 7.32. To obtain the diversity gain for the handheld employing selection combining at the 99.5% level, each value of branch power ratio and correlation must be mapped using Figure 7.31, resulting in the curves of Figure 7.35.

The diversity gain is nearly the same with and without the user present but the performance is not the same with and without the user. The diversity gain expresses the gain, at a certain level, relative to the best of the branches. Therefore, the performance can be found from the level of the strongest received power in each direction and the diversity gain. By adding the diversity gain to the strongest of the received powers in each direction, the effective MEG can be found. This is shown in Figure 7.36.

The effective MEG results are less directional compared to the MEG curves of the individual antennas. This is not due to the change in correlation or diversity gain but to the complementary MEG curves for the two antennas (when one is high, the other is low); see Figure 7.34. The gain due to the changes in MEG from the turning of the user is not included when diversity gain is usually considered,

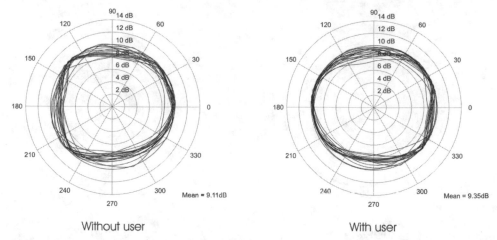

Figure 7.35   Diversity gain obtained by selection combining at the 99% level for the handheld equipped with a dipole and a patch antenna, as shown in Figure 7.33. (Left) Without a user and (right) with a model of the user. The handheld and the BS are located indoor in the same building. (From: [47]. ©1999 Union Radio-Scientifique Internationale [URSI]. Reprinted with permission.)

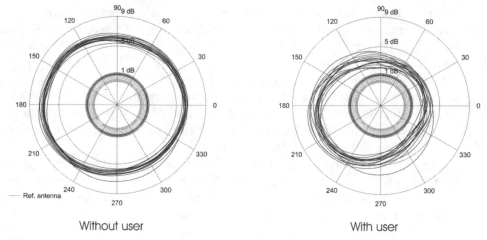

Figure 7.36   Effective MEG for selection combining at the 99% level for the dipole-patch configuration shown in Figure 7.33. As a reference the circles near 1 dB represent the MEG (without diversity) in the same environments for a vertical dipole (left) without a user and (right) with a model of the user. The handheld and the BS are located indoor in the same building. (From: [47]. ©1999 Union Radio-Scientifique Internationale [URSI]. Reprinted with permission.)

and this is an extra gain that corresponds to macro diversity. It can be obtained if the MEG curves are complementary (i.e., only one antenna has a notch in the MEG curve in a given direction.)

Some experimental results for two-branch diversity on a handset are given in [34] and [35], where it is verified that diversity gain can be achieved on a small handset, also in the presence of a human operator, owing to low correlation and similar MEG values for the two branches.

The expressions for the MEG in (7.7) and the correlation in (7.12) assume that the DUT remains within an area where the channel is stationary, so that the model of the environment is valid. As described in Section 7.4.1, the model of the environment is typically obtained from a number of measurements and represents the average distribution of power versus direction as received by the DUT. The drawback of this approach is that the distributions of received power cannot be directly obtained, making it difficult to estimate, for example, diversity gain, as shown above.

As an alternative to direct measurements of, for example, amplitude distributions of multiple receiver antennas in order to estimate the diversity gain, the authors of [36] suggest to use a method that may be considered a hybrid of using direct measurements and the ideas leading to the MEG expression in (7.7). Briefly, the method can be described as follows. Assuming that the instantaneous signal arriving at the DUT can be decomposed into a number of plane waves with known angles of arrival and amplitudes, and assuming the radiation patterns for all the ports of the DUT are known, then the instantaneous received signal can be computed using an expression similar to (7.3). Repeating this for all instantaneous channel descriptions, the complete sequence of received signals corresponding to the time-varying channel is obtained. By knowing the instantaneous outputs of all antenna ports, different combining methods may be used and the performance can be tested in the particular environment. Also other transmission methods such as MIMO may be investigated. The main drawback is that a valid plane wave description is needed that will often require multichannel channel sounding equipment and parameter estimation algorithms with sufficient resolution and accuracy.

7.5   Standard Measurement Setup Using Phantoms

For all modern systems of mobile communication where many vendors of both the infrastructure and the mobile phones deliver equipment, common interfaces are needed. One of the perhaps best specified interfaces is the radio interface where both the mobile phones as well as the base stations from different vendors need to communicate. This is ensured by a detailed description of its operation as well as a specification on how to test that the equipment fulfills the requirements. This is also the case for mobile phones today but nearly all tests on the phones are done at the RF terminal. The reason for using the RF connector and not the antenna output is purely practical. But as the phones have developed over time, very different designs of antennas, transceivers, and the phone cases have been made and the communication performance has varied significantly among the phones, as stated previously in this chapter [37, Sec. 3].

The variation in the performance among handsets is possible because there are so far no requirements in, for example, the GSM standard with respect to the actually transmitted and received power. Only power levels measured at the antenna terminals are specified. In an attempt to improve on this situation some work has been done in a working group of COST 259 and its successor COST 273 (European Co-operation in the Field of Scientific and Technical Research) [38]. This work has focused on performance evaluation based on measurements of the spherical radiation pattern of the handsets. Similarly, the Cellular Telecom-munications & Internet Association (CTIA) has been working on a certification of mobile handsets in terms of the TRP relevant for the UL and TIS for the DL [29]. As described in Section 7.4.1, the TRP and TIS may be seen as a special case of the MEG measure [1, 25]. Unlike TRP and TIS, the MEG takes into account both the directional and polarization properties of the handset antenna and the mobile environment.

As discussed in Section 7.3, the mobile performance is highly influenced by the presence of a user in a way that depends on how he or she uses the mobile. It is clear that test procedures need to include the typical user position that is very different for different types of user equipment (UE) such as phones, data termi-nals, video phones, arm wrist devices, pagers, and other types of UE that can be expected to come in the future. The most common UE type today is the phone and therefore the following focuses on the typical use of a phone next to the human head.

For practical reasons, measurements of the spherical radiation patterns usually do not include live test persons, as described in Section 7.3 and in [39]. Instead, the influence of the handset user on the performance is simulated by a phantom of the user's head that is placed next to the handset during measurements. An over-view of a practical setup for measurements of radiation patterns of mobile hand-sets is given in Section 7.5.1. An important aspect of carrying out measurements is the inevitable uncertainty, which is discussed in Section 7.5.2.

One source of uncertainty in the measurements is if the handset is not placed exactly as intended on the phantom. The importance of this is discussed further below in Section 7.5.5. An investigation into the influence of sampling density of the radiation patterns is described in Section 7.5.3, and the influence of the type of phantom used is described in Section 7.5.4.

7.5.1 Setup for Measurements of Mobile Handset Radiation Patterns

The task of measuring the radiation pattern of a mobile phone is not straight-forward because both the UL and DL have to be measured and because it must be done without attaching cables to the phone or otherwise modifying it in any way that may change the radiation patterns of the phone. An automated setup has been developed at Aalborg University that is sketched in Figure 7.37 and described below.

The measurements are conducted in an anechoic room and the setup is capa-ble of measuring both polarizations in any grid on a sphere. The anechoic room is large, $7 \times 7 \times 10$m, and the distance between the rotating phone and the probe antenna is some 5.5m. The measurements are based on a system tester, here a Rohde & Schwarz CMU200 that acts as the BS controlling the phone. In this manner no modifications are made to the phone that could disturb the radiation.

A measurement starts by the system tester initiating a call to the phone, which is answered on the phone by an operator, then the phone is fixed in the position for measurement and the program starts to measure the UL and DL in each polar-ization, after which the pedestal moves to the next position. After all desired posi-tions on the sphere are measured, the program moves to the position and polarization of maximum received power and starts reducing the power transmit-ted by the system tester in steps of 0.25 dB while the phone is reporting the

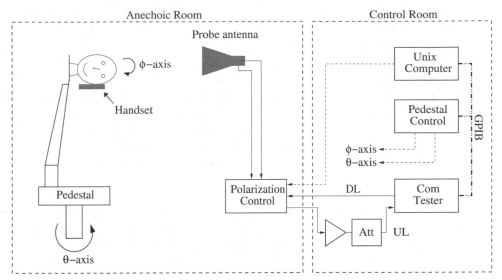

Figure 7.37   Overview of the measurement system. The measurements are made using different combinations of rotation about the vertical axis of the phantom and the pedestal. (From: [42]. ©2006 IEEE. Reprinted with permission.)

received power. In this way any deviation from the ideally linearly decreasing power is recorded, allowing for a compensation after the measurement of the power levels reported by the handset at DL.

For the UL, the measurements are based on the peak burst power including various system aspects, as the phone is an off-the-shelf product. The matching losses are not only the loss due to an SWR stemming from imperfect impedance matching but also the loss due to power amplifier load-pull. As no modifications are made to the phones, the measured radiated power is really the power available for communication and the radiation pattern itself is not altered due to external cables. Further, if limits are put on the minimum amount of radiated power for example, this measure will include all parameters influencing the transmitted power.

For the DL, the most accurate way to measure, as specified in the GSM specifications, is to measure the bit error rate (BER) and not directly the received power. As the measurements of BER are very time-consuming and since there is a direct relation between BER and received power (RxLev in GSM) for each individual phone, the RxLev is measured instead [40]. If the phone is measuring the received power well, no compensation for nonlinear behavior is needed, which is often the case.

The absolute power levels in both UL and DL are obtained using a back-to-back calibration of all cables, switches, splitters, amplifiers, and attenuators using a network analyzer. Calibration of the CMU200 is done by comparing to a peak power meter. The probe antenna is measured in a 3-antenna measurement setup to obtain the absolute gain, and the distance between the phone center (which is the center of rotation) and the probe antenna is used to calculate the losses by Friis' transmission law.

As a check of the calibration, a mobile phone was connected directly to a reference monopole antenna (one monopole antenna for 900 MHz and one for

1,800 MHz) and measured similarly to the phones in free space for each channel in both UL and DL. The power transmitted by the phone was measured and compared to the power obtained by integrating the transmitted and received power over the sphere. The loss in the reference monopole was some 0.5 dB for all frequencies used.

To establish some confidence in the measurements and to find the level of repeatability, two phones were measured 10 times in free space. Between measurements various changes were made such as measurements on fully charged battery compared to half-charged battery, and cold phone compared to phone that has transmitted on maximum power for 2 hours, as well as disassembling the setup, and so forth. The repeated measurements were interleaved with the other measurements throughout the time of all measurements for this investigation. The repeated measurements also showed a variation in both TRP and TIS of only ±0.13 dB. The radiation pattern was investigated through the MEG and an additional variation of less than ±0.16 dB was found for both UL and DL.

7.5.2 Measurement Uncertainty

When designing and specifying a setup for measuring the performance of mobile devices as the one described in the previous section, it is important to also consider the accuracy of resulting measurements. All involved equipment needs to be calibrated, either separately or combined, using either other already calibrated equipment or utilizing known properties of measured devices. For example, if the loss of a given antenna is known, it can be measured in the system that can then be calibrated using the acquired data. However, even a calibrated system only gives results within a certain margin due to issues such as measurement noise. Some examples of error sources in measurements are:

- Changed properties due to temperature differences at calibration time and measurement time;
- Minor variations in setup, such as different connectors;
- Reflections in anechoic room;
- Incorrect mounting of mobile device;
- Errors due to insufficient sampling of radiation pattern;
- Thermal noise.

The document [29] describes how mobile handsets must be measured in order to be certified by CTIA, and gives detailed instructions of how to handle the many uncertainties in a measurement system. Sections 7.5.3 to 7.5.5 describe in more detail some investigations carried out in order to find suitable compromises between measurement accuracy and speed/cost.

7.5.3 Errors Due to Insufficient Sampling Density of Radiation Pattern

In practice the surface integral of (7.7) involved in obtaining the MEG has to be computed from a finite set of samples of the spherical radiation pattern. Since the individual measurements are time consuming there is a trade-off between the accuracy and the time it takes to collect the measurements. Furthermore, if the radiation patterns are obtained from battery-powered handsets, there is a limit on the acceptable total measurement time. In addition, the radiation patterns are generally frequency-dependent and hence they may have to be measured at multiple frequencies. Therefore, some investigations are required as to how densely the radiation patterns need to be sampled for a given accuracy, and on how many frequencies.

The TRP may be seen as a special case of MEG and is obtained via a numerical integration, assuming an isotropic environment (see Section 7.2.4). Alternatively, instead of direct use of the measured radiation pattern samples in the numerical integration, the measured values may be used to estimate parameters of a so-called spherical wave expansion, describing the radiation pattern at arbitrary points on the sphere. The work described in [41] investigates the error in TRP obtained using different estimating methods, based on simulations of a helix and a patch on a mobile handset next to a phantom, consisting of a head, neck, and shoulders. Frequencies in the range 0.9 to 2.1 GHz were considered. It was found that only about 50 measurement directions are needed to estimate the correct TRP within 0.1 dB.

In practice the measurements in directions of the individual sampling points include errors due to issues such as reflections in the anechoic room, calibration errors, and measurement noise. These errors should also be included when investigating the number sampling points, since the influence of the errors for the individual sampling points may be reduced by the averaging inherent in the MEG measurement. When studying the effects of the sampling density it is important to distinguish between TRP/TIS values, corresponding to an isotropic environment with uniform weighting of the measured samples, and the MEG obtained with more realistic models of the environment with nonuniform weighting of the samples, both versus direction and for the two polarizations. A larger error is generally expected if nonuniform models are used.

An evaluation of the sampling density influence on the error in both general MEG values and TRP/TIS is described in [42], based on practical measurements on different mobile handsets and using different models of the environment. Applying numerical integration, as discussed above, a sampling density of $10° \times 20°$ in the θ and ϕ angles, respectively, resulted in a maximum error of 0.4 dB considering all environment models. If only the isotropic model is used to get TRP/TIS values a sampling density of $30° \times 30°$ could be used with a maximum error of 0.5 dB.

7.5.4   Investigation of Phantom Type Influence on Performance

In the choice of human phantom for performance measurements it is important to know to what degree different phantoms produce different results. This may have implications on how many phantoms and of what types of phantoms should be used in standardized measurements.

As an example, the work described next investigates the difference in using two phantoms. This work is based on measurements; other investigations using simulations into the influence of the phantom type are described in [43].

One of the phantoms considered here is the specific anthropomorphic mannequin (SAM) phantom, which models only the human head and neck and is based on head dimension statistics of American soldiers[44]. The phantom was origi-

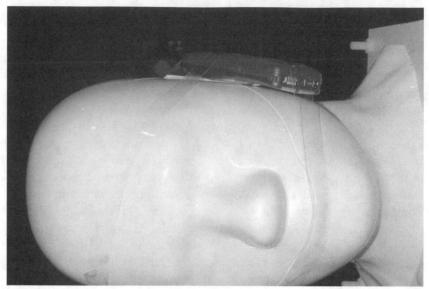

Figure 7.38   A phone during measurement on the SAM phantom in the anechoic room.

nally specified for SAR measurements, but has since been adopted by the CTIA for performance testing [29]. Figure 7.38 shows a phone on the SAM phantom during measurement. The other phantom models both the head, neck, and upper parts of the human chest, and is sold by Schmid & Partner [30]. This phantom is referred to below as the torso phantom (see Figure 7.39). In addition, all phones are also measured in free space for reference (see Figure 7.40). All measurements of the radiation patterns were done as described in Section 7.5.1 using a 10° × 10° sampling grid in the azimuth and elevation angles.

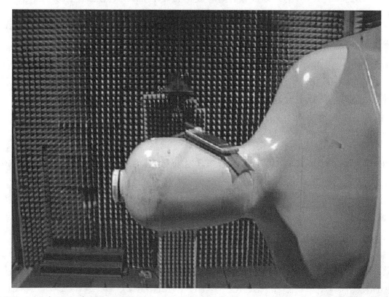

Figure 7.39   A phone during measurement on the torso phantom in the anechoic room.

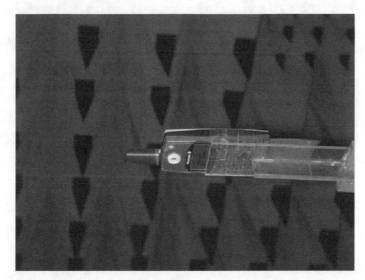

Figure 7.40   One of the phones during measurement in free space in the anechoic room.

The comparison is done in terms of TRP, TIS, and MEG (see Section 7.4) and for five typical and commercially available phones in 2002. The handsets were all dual-band GSM-900 and GSM-1800. The phones are the same as described in Section 7.4.2 (i.e., one large and one small phone with internal antenna, labeled B and C, respectively, and one large and one small phone with helical antennas, labeled A and F, in addition to a small phone with an extractable half-wave dipole antenna, labeled E.) The MEG results are obtained using the AAU and HUT models described in Section 7.4.1.

The results regarding TRP and TIS are given in Table 7.8 for the 900-MHz band and in Table 7.9 for the 1,800 MHz band. Comparing the two phantoms the difference in TRP and TIS is less than 0.6 dB for 900 MHz and less than 0.4 dB for 1,800 MHz, with one exception of 1.6 dB. The difference in MEG using the AAU model for the two phantoms is typically 0.6 dB and maximum 0.9 dB for 900 MHz, and typically 0.25 dB and maximum 0.7 dB for 1,800 MHz.

Table 7.8   TRP and TIS Values for Free Space, SAM, and the Torso Phantom at 900 MHz

Phone	Free Space		SAM		Torso	
	TRP	TIS	TRP	TIS	TRP	TIS
A	30.1	+0.1	25.7	−6.1	25.3	−6.6
B	31.2	−0.6	26.2	−6.5	26.0	−7.0
C	30.7	−1.5	25.6	−7.2	25.2	−7.8
E	27.5	−2.8	26.4	−4.5	26.4	−4.6
F	30.2	−3.0	26.2	−8.4	26.1	−8.9

The TIS values are with respect to the reference TIS.

Source: [49].

Table 7.9   TRP and TIS values for Free Space, SAM, and the Torso Phantom at 1,800 MHz

Phone	Free Space		SAM		Torso	
	TRP	TIS	TRP	TIS	TRP	TIS
A	26.4	−2.5	23.6	−5.5	23.2	−5.7
B	27.7	−3.8	24.4	−6.9	24.2	−7.2
C	28.0	−1.5	26.1	−4.8	26.2	−3.2
E	26.8	−3.1	23.9	−4.7	24.2	−4.8
F	26.3	−2.0	24.4	−4.8	24.4	−4.9

The TIS values are with respect to the reference TIS.

Source: [49].

7.5.5   Uncertainty Due to Incorrect Handset Positioning on Phantom

Relating to the discussion in Section 7.5.2 it is important to know how much an inaccurate positioning of the mobile handset on the phantom during measurement of the radiation pattern will affect the results in terms of TRP, TIS, and MEG. With the aim of quantifying this influence, a series of spherical radiation pattern measurements were carried out at Aalborg University, where all handsets were measured in both the reference position on the phantom as well as in several slightly changed (i.e., incorrect) positions. The results of these investigations are presented next.

The handsets involved in this investigation are listed in Table 7.10, which includes some of the same handsets as used in Section 7.5.4 in addition to Handset G and H. The set of handsets represents some of the most important types available on the market. The measurements were made using the system described in Section 7.5.1 and spherical radiation patterns were sampled using increments of 10° in the elevation angle θ and 20° in the azimuth angle ϕ. The reason for the more dense sampling in the elevation angle is that the dimensions of the combined handset and phantom are larger along the elevation angle than the azimuth angle. In [42] this choice of sampling density has been shown to lead to negligible errors. All results, however, are based on radiation patterns sampled in a 15° × 15° grid, obtained via interpolation of the measured data. This was done in order to meet the requirements of the CTIA certification document [29]. The interpolation is needed in any case to obtain rotated radiation patterns (see below), since samples are needed from directions not in the original sampling grid. The spline interpolation method was used. The handsets were measured next to a SAM phantom head [44], which was filled with a tissue simulating liquid as required by the CTIA certification.

In assessing the changes in the MEG and TRP/TIS values the repeatability of the measurement procedure itself must be known. In a similar campaign carried out in the same anechoic room, this was investigated using repeated measurements of the radiation patterns. The measurements included dismounting and mounting of the handset and the MEG results were found typically to be repeatable within 0.1 to 0.3 dB [42].

Table 7.10   Overview of the Measured Handsets

Handset	H × W × D	Antenna Type	Ant. Dist. to Front	Handset Type
A	130×47×23	External	19	Candybar
B	129×47×18	Internal	18	Candybar
E	97×50×15	Whip	11	Flip
F	97×50×15	Helix	11	Flip
G	88×50×19	External	13	Clamshell
H	96×43×19	Internal	19	Candybar

Distances are given in mm.

The clamshell handset G is measured in closed condition. When opened, the antenna is located near the joint of the two halves.

Source: [45].

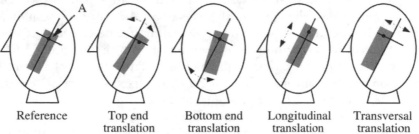

Reference   Top end   Bottom end   Longitudinal   Transversal
                translation   translation   translation   translation

Figure 7.41   Handset positions on the phantom. Only one direction is shown for each type of translation. (Reprinted from [45] with permission from Springer-Verlag GmbH.)

During measurements the handset was mounted on the left side of the phantom, similar to Figure 7.38. Five different measurement series were made, each differing in the way the handset was mounted on the head, as given below and sketched in Figure 7.41. In all cases the handsets were mounted on the phantom using Teflon tape. The terminology used for specifying the handset position was adopted from the CTIA certification document.

- *Reference.* In this reference measurement series, the handset was mounted according to the CTIA document [29].

- *Top end translation.* In this series, the handsets were mounted in the reference position except that the A point of the handset was rotated about the bottom (the point touching the phantom) so that the A point was ±15 mm off the correct position, where the distance was the position of the A point projected on the horizontal reference line of the handset, and where the negative offset was toward the face of the phantom.

- *Bottom end translation.* Similar to the top translation series, this series rotated the bottom reference point about the ear reference point. The distance was measured along the line passing through the lower reference point and was perpendicular to the vertical center line of the handset.

- *Longitudinal translation.* In this series of measurements the handset was moved along the vertical handset center line so that the A point was either above (+15 mm) or below (–7.5 mm) the ear reference point. The negative value was chosen to be –7.5 mm rather than –15 mm because handset E/F cannot rest on the phantom ear in a reasonable way if the larger translation is used, and thus this mounting is unrealistic.

- *Transversal translation.* In this series of measurements the handset was translated either toward the face of the phantom (–15 mm) or toward the back (+15 mm) while the handset center line was kept parallel to the line connecting the mouth and ear reference points (the MB line).

The translation distances used were deliberately chosen rather large in order to create a worst-case scenario. If the handset is carefully mounted on the phantom the translations will be smaller in practice.

For all handsets in actual use both the radiation pattern and the spherical power distribution are directive, and the MEG will vary depending on the orientation of the handset with respect to the environment.

In order to investigate this, the measured radiation patterns were rotated using two angles as described previously in Section 7.4.2. For each combination of rotation angles the MEG was computed. Note that the described postprocessing rotation procedure corresponds to a rotation of both the handset and the phantom. Thus, this is not a rotation of the handset relative to the phantom, but rather a rotation of the phantom with the handset at a fixed angle relative to the phantom. Evaluation of the MEG for different rotations of the handset relative to the phantom requires measurement of the radiation pattern for each rotation angle. This was not done in this work since it would result in a large number of measurements. Furthermore, only small differences are expected comparing the MEG computed from data obtained via a rotation, using the postprocessing procedure described above, and the MEG obtained using measurements of the radiation pattern obtained with the handset fixed at the desired angle on the phantom.

The MEG values were computed from the discrete equivalent of (7.7) and using the AAU, HUT, and ISO models described in Section 7.4.1. In the following the change in MEG due to the various translations has been investigated using the normalized MEG defined as

$$\Gamma'\left(\lambda,\mu\right)=\frac{\Gamma\left(\lambda,\mu\right)}{\Gamma_{ref}\left(\lambda,\mu\right)} \tag{7.13}$$

where $\Gamma(\lambda,\mu)$ is the MEG for a specific radiation pattern measurement and using the rotation angles (λ,μ), and $\Gamma_{ref}(\lambda,\mu)$ is the corresponding reference measurement for the same handset. Thus, for each measurement a large number of values are obtained, one for each orientation of the handset. For this reason the mean and standard deviation have been used for the analysis, except for the isotropic environment, which yields the same value irrespective of the handset orientation.

Figure 7.42 shows the change in TRP and TIS for each handset and frequency band, grouped in the different types of translations from the reference position. The different combinations of handsets and translations are shown

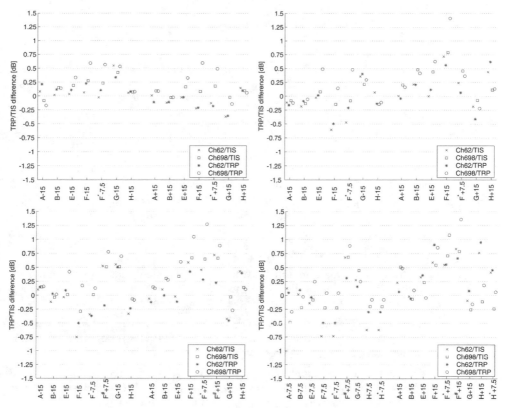

Figure 7.42   Change in TRP and TIS for the different types of translation. Translation of bottom end (top left), translation of top end (top right), transversal translation (bottom left), and longitudinal translation (bottom left). (Reprinted from [45] with permission from Springer-Verlag GmbH.)

along the *x*-axis where, for example, E-15 means handset E translated −15 mm. The data for handsets marked with a '*' or '#' are repeated or extra measurements. Generally it was found that TIS and TRP values are correlated so that, for example, an increase of the transmitted power due to a translation is usually associated with an increase in the received power. Furthermore, the results show similar effects for the low and high frequency bands. The deviations found for the TIS and TRP values are generally within ±0.5 dB with a maximum deviation of about 1.4 dB. From statistics of the computed MEG values based on data from all handsets, link directions, orientation, and offsets, it is found that the mean MEG deviations due to translations are generally low, about 0 to 0.2 dB. Furthermore, standard deviations of 0.1 to 0.5 dB and maximum deviations up to 1.6 dB were found for most handsets, with one exception having a maximum up to 2 dB. More results and details are available in [45].

The changes due to the incorrect position of the handsets on the phantom should be compared to the uncertainty due to the measurement system and the methods used. Using repeated measurements, the MEG results were found typically to be repeatable within 0.1 to 0.3 dB in [42]. In addition, the changes in the MEG introduced by positioning errors should be compared to the variation in the MEG of 6 to 8 dB that may be observed for a handset depending on its general orientation in the environment. For the TRP/TIS a difference of 3 to 4 dB was found between different types of handsets [46].

References

[1]   Andersen, J. B., and F. Hansen, "Antennas for VHF/UHF Personal Radio: A Theoretical and Experimental Study of Characteristics and Performance," *IEEE Transactions on Vehicular Technology*, Vol. 26, No. 4, November 1977, pp. 349–357.

[2]   Nielsen, J. Ø., and G. F. Pedersen, "In-Network Performance of Handheld Mobile Terminals," *IEEE Transactions on Vehicular Technology*, Vol. 55, No. 3, 2006, pp. 903–916.

[3]   Nielsen, J. Ø., G. F. Pedersen, and C. Solis, "In-Network Evaluation of Mobile Handset Performance," *Proc. IEEE Vehicular Technology Conference*, Vol. 2, pp. 732–739.

[4]   Standard definitions of terms for antennas, revised IEEE Std. 145-1993, http://ieeexplore.ieee.org/servlet/opac?punumber=2785, June 1993.

[5]   Wheeler, H. A., "The Radiansphere Around a Small Antenna," *Proc. of the IRE*, Vol. 47, No. 8, 1959, pp. 1325–1331.

[6]   Ohnishi, N., K. Sasaki, and H. Arai, "Field Simulator for Testing Handset Under Multi-Path Propagation Environments," *Proc. IEEE Antennas and Propagation Society International Symposium, Vol. 4,* 1997, pp. 2584–2587.

[7]   Rosengren, K., et al., "A New Method to Measure Radiation Efficiency of Terminal Antennas," *Proc. IEEE Conference on Antennas and Propagation for Wireless Communications*, 2000, pp. 5–8.

[8]   Krogerus, J., K. Kiesi, and V. Santomaa, "Evaluation of Three Methods for Measuring Total Radiated Power of Handset Antennas," *Proc. IEEE Instrumentation and Measurement Technology Conference*, Vol. 2, 2001, pp. 1005–1010.

[9]   Icheln, C., J. Krogerus, and P. Vainikainen, "Use of Balun Chokes in Small-Antenna Radiation Measurements," *IEEE Transactions on Instrumentation and Measurement*, Vol. 53, No. 2, 2004, pp. 498–506.

[10]   Icheln, C., J. Ollikainen, and P. Vainikainen, "Reducing the Influence of Feed Cables on Small Antenna Measurements," *Electronics Letters*, Vol. 35, No. 15, July 1999, pp. 1212–1214.

[11]   Kotterman, W. A. Th., G. F. Pedersen, and P. Eggers, "Cable-Less Measurement Set-Up for Wireless Handheld Terminals," *Proc. Personal, Indoor and Mobile Radio Communications Conference,* September 2001, pp. B112–B116.

[12]   Arai, H., N. Igi, and H. Hanaoka, "Antenna-Gain Measurement of Handheld Terminals at 900 MHz," *IEEE Transactions on Vehicular Technology*, Vol. 46, No. 3, August 1997, pp. 537–543.

[13]   Murase, M., Y. Tanaka, and H. Arai, "Propagation and Antenna Measurements Using Antenna Switching and Random Field Measurements," *IEEE Transactions on Vehicular Technology*, Vol. 43, No. 3, August 1994, pp. 537–541.

[14]   Dimbylow, P. J., and S. M. Mann, "SAR Calculations in an Anatomically Realistic Model of the Head for Mobile Communication Transceivers at 900 MHz and 1.8 GHz," *Physics in Medicine and Biology*, Vol. 39, No. 10, 1994, pp. 1537–1553.

[15]   Jensen, M. A., and, Y. Rahmat-Samii, "EM Interaction of Handset Antennas and a Human in Personal Communications," *Proc of the IEEE*, Vol. 83, No. 1, January 1995, pp. 7–17.

[16]   Tay, R. Y. -S., Q. Balzano, and N. Kuster, "Dipole Configurations with Strongly Improved Radiation Efficiency for Hand-Held Transceivers," *IEEE Transactions on Antennas and Propagation*, Vol. 46, No. 6, June 1998, pp. 798–806.

[17]   Toftgard, J., S. N. Hornsleth, and J. B. Andersen, "Effects on Portable Antennas of the Presence of a Person," *IEEE Transactions on Antennas and Propagation*, Vol. 41, No. 6, 1993, pp. 739–746.

[18]   Boyle, K., "Mobile Phone Antenna Performance in the Presence of People and Phantoms," *Proc. IEE Antennas and Propagation Professional Network Technical Seminar on Antenna Measurement and SAR*, May 2002, pp. 8/1–8/4.

[19]   Nielsen, J. Ø., et al., "Statistics of Measured Body Loss for Mobile Phones," *IEEE Transactions on Antennas and Propagation*, Vol. 49, No. 9, September 2001, pp. 1351–1353.

[20]   Pedersen, G. F., M. Tartiere, and M. B. Knudsen, "Radiation Efficiency of Handheld Phones," *Proc. IEEE Vehicular Technology Conference*, Vol. 2, May 2000, pp. 1382–1385.

[21]   Pedersen, G. F., et al., "Measured Variation in Performance of Handheld Antennas for a Large Number of Test Persons," *Proc. IEEE Vehicular Technology Conference*, May 1998, pp. 505–509.

[22]   Graffin, J., N. Rots, and G. F. Pedersen, "Radiation Phantoms for Handheld Phones," *Proc. IEEE Vehicular Technology Conference*, Vol. 2, 2000, pp. 853–866.

[23]   Krogerus, J., et al., "Effect of the Human Body on Total Radiated Power and the 3-D Radiation Pattern of Mobile Handsets," *IEEE Transactions on Instrumentation and Measurement*, Vol. 56, No. 6, December 2007, pp. 2375–2385.

[24]   Jakes, W. C., (ed.), *Microwave Mobile Communications*, New York: IEEE Press, 1974.

[25]   Taga, T., "Analysis for Mean Effective Gain of Mobile Antennas in Land Mobile Radio Environments," *IEEE Transactions on Vehicular Technology*, Vol. 39, No. 2, May 1990, pp. 117–131.

[26]   Nielsen, J. Ø., et al., "Computation of Mean Effective Gain from 3D Measurements," *Proc. IEEE Vehicular Technology Conference*, Vol. 1, July 1999, pp. 787–791.

[27]   Knudsen, M. B., and G. F. Pedersen, "Spherical Outdoor to Indoor Power Spectrum Model at the Mobile Terminal," *IEEE Journal on Selected Areas in Communications*, Vol. 20, No. 6, August 2002, pp. 1156–1169.

[28]   Kalliola, K., et al., "Angular Power Distribution and Mean Effective Gain of Mobile Antenna in Different Propagation Environments," *IEEE Transactions on Vehicular Technology*, Vol. 51, No. 5, September 2002, pp. 823–838.

[29]   Cellular Telecommunications & Internet Association (CTIA), *CTIA Test Plan for Mobile Station over the Air Performance*, Revision 2.0. Technical Report, CTIA, March 2003, http://www.ctia.org.

[30]   Schmid & Partner, http://www.speag.com/, Generic torso phantom, v.3.6.

[31]   Schwartz, M., W. R. Bennet, and S. Stein, *Communication Systems and Techniques*, New York: McGraw-Hill, 1966.

[32]   Vaughan, R. G., and J. B. Andersen, "Antenna Diversity in Mobile Communications," *IEEE Transactions on Vehicular Technology*, Vol. 36, No. 4, November 1987, pp. 149–172.

[33]   Risom, J. B., and L. Grevenkop-Castenskiold, "Antenna Diversity for DECT Portable Part," Master's Thesis, Aalborg University, Department of Electronic Systems, 1997.

[34]   Green, B. M., and M. A. Jensen, "Diversity Performance of Dual-Antenna Handsets Near Operator Tissue," *IEEE Transactions on Antennas and Propagation*, Vol. 48, No. 7, July 2000, pp. 1017–1024.

[35]   Ogawa, K., T. Matsuyoshi, and K. Monma, "An Analysis of the Performance of a Handset Diversity Antenna Influenced by Head, Hand, and Shoulder Effects at 900

MHz: Part II–Correlation Characteristics," *IEEE Transactions on Vehicular Technology*, Vol. 50, No. 3, May 2001, pp. 845–853.

[36]   Suvikunnas, P., et al., "Evaluation of the Performance of Multiantenna Terminals Using a New Approach," *IEEE Transactions on Instrumentation and Measurement*, Vol. 55, No. 5, 2006, pp. 1804–1813.

[37]   Correia, L. M., (ed.), *Wireless Flexible Personalised Communications. COST 259: European Co-operation in Mobile Radio Research*, New York: Wiley, 2001.

[38]   *Co-Operation in the Field of Scientific and Technical Research* (COST), http://www.lx.it.pt/cost273/.

[39]   Pedersen, G. F., K. Olesen, and S. L. Larsen, "Bodyloss for Handheld Phones," *Proc. IEEE Vehicular Technology Conference*, Vol. 2, July 1999, pp. 1580–1584.

[40]   Knudsen, M. B., "Antenna Systems for Handsets," Ph.D. Thesis, Center for PersonKommunikation, Aalborg University, September 2001, http://home1.stofanet.dk/grenen7.

[41]   Laitinen, T. A., et al., "Amplitude-Only and Complex Field Measurements for Characterizing Radiated Fields of Mobile Terminal Antennas from a Small Number of Samples," *IEEE Transactions on Instrumentation and Measurement*, Vol. 54, No. 5, October 2005, pp. 1989–1996.

[42]   Nielsen, J. Ø., and G. F. Pedersen, "Mobile Handset Performance Evaluation Using Radiation Pattern Measurements," *IEEE Transactions on Antennas and Propagation*, Vol. 54, No. 7, 2006, pp. 2154–2165.

[43]   Ogawa, K., T. Matsuyoshi, and K. Monma, "An Analysis of the Performance of a Handset Diversity Antenna Influenced by Head, Hand, and Shoulder Effects at 900 MHz: Part I–Effective Gain Characteristics," *IEEE Transactions on Vehicular Technology*, Vol. 50, No. 3, May 2001, pp. 830–844.

[44]   *Specific Anthropomorphic Mannequin (SAM) Phantom*, http://www.sam-phantom.com/.

[45]   Nielsen, J. Ø., and G. F. Pedersen, "Influence of Spherical Radiation Pattern Measurement Uncertainty on Handset Performance Measures," *Wireless Personal Communications*, Vol. 32, No. 1, January 2005, pp. 9–22.

[46]   Nielsen, J. Ø., and G. F. Pedersen, "Mobile Handset Performance Evaluation Using Spherical Measurements," *Proc. IEEE Vehicular Technology Conference*, Vol. 1, September 2002, pp. 289–293.

[47]   Pedersen, G. F., and J. B. Andersen, "Handset Antennas for Mobile Communications–Integration, Diversity, and Performance," W. R. Stone (ed.), *Review of Radio Science 1996–1999*, New York: Oxford University Press, 1999, pp. 119–137.

[48]   Pedersen, G. F., and S. Skjærris, "Influence on Antenna Diversity for a Handheld Phone by the Presence of a Person," *Proc. IEEE Vehicular Technology Conference*, 1997, pp. 1768–1772.

[49]   Pedersen, G. F., and J. Ø. Nielsen, "Radiation Pattern Measurements of Mobile Phones Next to Different Head Phantoms," *Proc. IEEE Vehicular Technology Conference*, 2002, pp. 2465–2469.

List of Acronyms

1G	first generation
2G	second generation
3D	three dimensional
3G	third generation
4G	fourth generation
AC	alternating current
ADG	apparent diversity gain
AMPS	Advanced/Analog Mobile Phone System
AUT	antenna under test
BER	bit error rate
BWR	bandwidth ratio
BS	base station
CDF	cumulative probability density function
CDMA	Code Division Multiple Access
CENELEC	European Committee for Electrotechnical Standardization
CF	constraint factor
CI	cochlear implant
COST	European Co-operation in the Field of Scientific and Technical Research
CPW	coplanar waveguide
CST	computer simulation technologies
CTIA	Cellular Telecommunications & Internet Association
CxP	cross-coupling
DA	averaged pointwise dimension
DAMPS	Digital Advanced Mobile Phone System
DB	box counting dimension
DC	direct current/correlation dimension
DCS	digital cellular system
DD	divider dimension
DE	Euclidean dimension
DECT	Digital Enhanced Cordless Telephone System
DGL	diversity gain loss
DH	Hausdorff dimension
DI	information dimension
DL	Lyapunov dimension

DL	downlink
DP	pointwise dimension
DS	similarity dimension
DUT	device under test
DT	topological dimension
DVB-H	digital video broadcasting handheld
EDG	effective diversity gain
E-GSM	extended GSM
EIRP	equivalent isotropic radiated power
EM	electromagnetic
EMC	electromagnetic compatibility
EMEG	effective mean effective gain
FBW	fractional matched VSWR bandwidth
FCC	Federal Communications Commission
FDD	frequency division duplex
FDTD	finite difference time domain
FOMA	freedom of mobile multimedia access
FR4	flame retardant 4
GSM	Global System for Mobile Communications
GSM-R	railways GSM
GPS	Global Positioning System
HA	hearing aid
HAC	hearing aid compatible handset
HFSS	high frequency structure simulator
HiperLAN2	high performance radio LAN 2
HSL	head simulating liquid
IDG	ideal diversity gain
IEEE	Institute of Electrical and Electronic Engineers
IET	Institution of Engineering and Technology
IFA	inverted-F antenna
IL	inverted-L
ILA	inverted-L antenna
IMT-2000	International Mobile Telecommunications 2000
ISM	industrial, scientific, and medical
ITU	International Telecommunication Union
LORAN	long-range navigation
LoS	line of sight
MCL	Microwave Consultants Limited
MIMO	multiple input multiple output
MEG	mean effective gain
MEM	microelectromechanical
MMIC	monolithic microwave integrated circuit
MoM	method of moments
MPIE	mixed potential integral equation
MRC	maximum ratio combining
MS	mobile station

MS	modal significance
NLoS	nonline of sight
NTT	Nippon Telegraph and Telephone
OFDM	orthogonal frequency division multiplexing
PCB	printed circuit board
PCS	personal communications service/system
PDC	personal digital cellular
P-GSM	primary GSM
PIFA	planar inverted-F antenna
PIL	parasitic inverted-L
PILA	planar inverted-L antenna
PSO	particle swarm optimization
RLC	resistance, inductance, and capacitance
RF	radio frequency
RWG	Rao-Wilton-Glisson
Rx	receiver
SAM	specific anthropomorphic mannequin
SAR	specific absorption rate
SC	selection combining
SC-FDMA	single-carrier frequency division multiple access
SDMB	satellite digital media broadcast
SER	symbol error rate
SNR	signal-to-noise ratio
STP	space-time processing
SWR	standing wave ratio
TACS	total access communication system
TDMA	time division multiple access
TEM	transversal electromagnetic
TIS	total isotropic sensitivity
TLM	transmission-line matrix
T-GSM	TETRA GSM
TPD	true polarization diversity
TRP	total radiated power
Tx	transmitter
UE	user equipment
UL	uplink
UMTS	Universal Mobile Telephone System
UNII	Unlicensed National Information Infrastructure
USA	United States of America
UTRA	UMTS Terrestrial Radio Access
UWB	ultrawideband
VSWR	voltage standing-wave ratio
WCDMA	wideband code division multiple access
Wi-Fi	wireless fidelity
WLAN	wireless local area network
WLL	wireless local loop

Wi-Max	wideband metropolitan area network
W-OFDM	wideband orthogonal frequency division multiplexing
XPD	cross-polarization discrimination
XPR	cross-polarization ratio

About the Editor

David A. Sánchez-Hernández is with Universidad Politécnica de Cartagena, Spain, where he leads the microwave, radio-communications and electromagnetism engineering research group (GIMRE, http://www.gimre.upct.es). He is the author of over 40 scientific papers, 80 conference contributions, 9 patents, and 8 books, acting regularly as a reviewer of many IET and IEEE publications and conferences. He has also been awarded several national and international prizes, including the R&D J. Langham Thompson Premium, awarded by the Institution of Electrical Engineers (formerly The Institution of Engineering and Technology, IET), or the *i-patentes* award by the Spanish Autonomous Region of Murcia to innovation and technology transfer excellence. He is also an IET Fellow Member and IEEE Senior Member. His research interests encompass all aspects of the design and application of printed multiband antennas for mobile communications, electromagnetic dosimetry issues, and MIMO techniques. He is the cofounder of EMITE Ingenieria SLNE.

About the Authors

Eva Antonino-Daviú

Eva Antonino-Daviú received the M.S. and Ph.D. degrees in electrical engineering from the Universidad Politécnica de Valencia, Spain in 2002 and 2008, respectively. In 2005 she stayed for several months as a guest researcher at IMST GmbH, in Kamp-Lintfort, Germany. Since 2005, she has been a lecturer at the Escuela Politécnica Superior de Gandía, Gandía, Spain. She is a member of the IEEE. Her research interests include wideband and multi-band planar antenna design and optimization and computational methods for printed structures.

Steven R. Best

Steven R. Best received B.Sc.Eng. and Ph.D. degrees in electrical engineering from the University of New Brunswick, Canada, in 1983 and 1988, respectively. He is currently a principal sensor systems engineer with the MITRE Corporation in Bedford, Massachusetts, where he is involved in supporting a number of government programs. He is the author or coauthor of over 100 papers in various journal, conference, and industry publications. He has served as a distinguished lecturer for the IEEE Antennas and Propagation Society and an associate editor for *IEEE Antennas and Wireless Propagation Letters*. He is an associate editor of the *IEEE Transactions on Antennas and Propagation,* a member of the IEEE APS AdCom, and the Junior Past Chair of the IEEE Boston Section. He is a Fellow of the IEEE and a member of ACES.

Kevin R. Boyle

Kevin R. Boyle received a B.Sc. (Hons.) in electrical and electronic engineering from City University, London, in 1989, an M.Sc. (with distinction) in microwaves and optoelectronics from University College, London, and a doctor of technology degree from Delft University of Technology in 2004. He was with Marconi Communications Systems Ltd., Chelmsford, England, until 1997, working on all aspects of antenna system design. He joined Philips Research Laboratories, Redhill, England, in 1997 (which became NXP Semiconductors Research in 2006), where he is currently a principal research scientist and a project leader for antenna and propagation related activities. His areas of interest include antenna design for mobile communication systems, diversity, propagation modeling, and related areas of mobile system design. Dr. Boyle has actively participated in COST 259 and COST 273, is a member of the IEEE, IET, and he is a chartered engineer.

He has published more than 25 papers in refereed international journals and conferences and holds more than 15 patents.

Marta Cabedo-Fabrés

Marta Cabedo-Fabrés received an M.S. degree in electrical engineering from the Universidad Politécnica de Valencia, Spain, in 2001, and a Ph.D. degree in electrical engineering at the same university in 2007. In 2001, she joined the Electromagnetic Radiation Group at Universidad Politécnica de Valencia (UPV) as a research assistant. In 2004 she became an associate professor at the Communications Department, Universidad Politécnica de Valencia. She is a member of the IEEE. Her current scientific interests include numerical methods for solving electromagnetic problems and design and optimization techniques for wideband and multiband antennas.

Wojciech J. Krzysztofik

Wojciech J. Krzysztofik received M.Sc. and Ph.D. degrees in electrical engineering from Wroclaw University of Technology, Wroclaw, Poland, in 1974 and 1983, respectively. Since 1974 he has been a member of the Department of Electronics Engineering, the Institute of Telecommunications & Acoustics of the Wroclaw University of Technology, where he is currently an assistant professor. Dr. Krzysztofik was a vice dean of the Faculty of Electronics Engineering at WUT of Wroclaw from 2002 to 2005. He has visited different research institutes, universities, and industries in Europe and the United States. In 1988 he was a visiting professor at the Chalmers University of Technology, Sweden. From 1988 to 1990 he was engaged as a professor of telecommunications at the College of Engineering of Mosul University, Iraq. In 2002 he visited the three U.S. universities in Texas: the University of Texas at Austin, University of Houston at Houston, and the Texas A&M University at College Station, presenting lectures and seminaries. His research interests are in microwave remote sensing microwave application in communication, the electromagnetic theory and numerical methods applied to scatterers, and antennas of arbitrary shape in complex environment, and he has published over 100 papers in these areas in journals and conference proceedings. He was a manager of the scientific research project entitled Investigation of Satellite Signal Reception in Poland, supported by the Polish Council of Scientific Research. Dr. Krzysztofik is a member of the WUT scientific group engaged in European COST projects: High Altitude Platforms for Communications and Other Services, and Innovative Antennas for Emerging Terrestrial & Space-Based Applications (under organization). In 2006 he joined the European Antenna Centre of Excellence in Nice, France. Dr. Krzysztofik is a member of Committee of Electronics and Telecommunication at the Polish Academy of Science (KEiT PAN), the Polish Society of Electrical Engineers (SEP), and the Polish Association of Theoretical & Applied Electro-Techniques (PTETiS). He is also a Senior Member of the IEEE AP/AES/MTT Joint Chapter, the Polish Section IEEE, where he held the position of vice chair of this chapter from 1998 to 2002, and he currently chairs the Polish joint chapter for incoming cadence 2007–2008. In 2002, Dr. Krzysztofik was elected as the chairperson of the Wroclaw Regional Seminary of the KEiT PAN/IEEE PL Joint Chapter AP-AES-MTT on Antennas, Microwaves &

Mobile Communications. He is on the board of reviewers of several international journals, including the *IEEE Transactions on Antennas and Propagation* and *IEE Proceedings Microwaves, Antennas and Propagation.*

RongLin Li

RongLin Li received a B.S. degree in electrical engineering from Xi'an Jiaotong University, China, in 1983, and M.S. and Ph.D. degrees in electrical engineering from Chongqing University, in 1990 and 1994, respectively. From 1983 to 1987, he worked as an electrical engineer at Yunnan Electric Power Research Institute. From 1994 to 1996, he was a postdoctoral research fellow in Zhejiang University, China. In 1997, he was with Hosei University, Japan, as an HIF (Hosei International Fund) research fellow. Since 1998, he has become a professor in Zhejiang University. In 1999, he served as a research associate at University of Utah. In 2000, he was a research fellow in Queen's University of Belfast, United Kingom. Since 2001, he has been with the ATHENA group as a research scientist at Georgia Institute of Technology. His latest research interests include computational electromagnetics, modeling of antennas and microwave devices, and RF packaging design.

Antonio M. Martínez-González

Antonio M. Martínez-González obtained his Dipl.-Ing. in telecommunications engineering from Universidad Politécnica de Valencia, Spain, in 1998, and his Ph.D. from Universidad Politécnica de Cartagena in 2004. From 1998 to 1999, he was employed as technical engineer at the Electromagnetic Compatibility Laboratory of Universidad Politécnica de Valencia, where he developed assessment activities and compliance certifications with European directives related with immunity and emissions to electromagnetic radiation from diverse electrical, electronic, and telecommunication equipment. He has been an assistant lecturer at Universidad Politécnica de Cartagena since 1999. In 1999 he was awarded the Spanish National Prize from Foundation Airtel and Colegio Oficial de Ingenieros de Telecomunicación de España for the best final project on mobile communications. His research interest is focused on electromagnetic dosimetry and radioelectric emissions and MIMO techniques for wireless communications.

Marta Martínez-Vázquez

Marta Martínez-Vázquez obtained a Dipl.-Ing. in telecommunications and a Ph.D. from Universidad Politécnica de Valencia, Spain, in 1997 and 2003, respectively. In 1999 she obtained a fellowship from the Pedro Barrié de la Maza Foundation for postgraduate research at IMST GmbH, in Germany. Since 2000, she has been a full-time staff member of the Antennas and EM Modelling Department of IMST. Her research interests include the design and applications of antennas for mobile communications, planar arrays and radar sensors, and electromagnetic bandgap (EBG) materials. Dr. Martínez-Vázquez was awarded the 2004 Premio Extraordinario de Tesis Doctoral (Best Ph.D. award) of the Universidad Politécnica de Valencia for her dissertation on small multiband antennas for handheld terminals. She is a member of the IEEE and a distinguished lecturer for the Antennas & Propagation Society of the IEEE. She has been a member of the Executive

Board of the ACE (Antennas Centre of Excellence) Network of Excellence (2004–2007) and the leader of its small antenna activity. She is the vice chair of the COST IC0306 Action Antenna Sensors and Systems for Information Society Technologies, and a member of the Technical Advisory Panel for the Antennas and Propagation Professional Network of IET. She is the author of over 50 papers in journals and conference proceedings.

Jesper Ødum Nielsen

Jesper Ødum Nielsen received his master's degree in electronics engineering in 1994 and a Ph.D. in 1997, both from Aalborg University, Denmark. He is currently employed at Department of Electronic Systems at Aalborg University where his main areas of interests are experimental investigation of the mobile radio channel and the influence on the channel by mobile users. He has been involved in channel sounding and modeling, as well as measurements using the live GSM network. In addition, he has recently been working with handset performance evaluation based on spherical measurements of handset radiation patterns and power distribution in the mobile environment.

Bo Pan

Bo Pan received B.S. and M.S. degrees from Tsinghua University, Beijing, China, in 2000 and 2003 respectively. He worked as a visiting student at Microsoft Research Asia before he started his Ph.D. study in 2003 in the school of electrical and computer engineering at the Georgia Institute of Technology, Atlanta. He was expecting to receive his Ph.D. in 2008. He is a research member of Georgia Electronic Design Center (GEDC) and GT-NSF Packaging Research Center (PRC). His research involves the design and fabrication of antennas for multiple wireless applications design, as well as development and characterization of various components and circuits for RF/microwave/millimeter-wave T/R modules, with the focus on micromachining technologies. He has authored and coauthored more than 20 papers in referred journals and conference proceedings.

Gert Frølund Pedersen

Gert Frølund Pedersen received a B.Sc.E.E. with honours in electrical engineering from the College of Technology, Dublin, Ireland, in 1991, and M.Sc.E.E. and Ph.D. degrees from Aalborg University, Denmark, in 1993 and 2003, respectively. He has been employed by Aalborg University since 1993, where he currently is working as a professor heading the Antenna and Propagation Group. His research has focused on radio communication for mobile terminals including small antennas, antenna systems, propagation, and biological effects. He has also worked as a consultant for development of antennas for mobile terminals, including the first internal antenna for mobile phones in 1994 with very low SAR, the first internal triple-band antenna in 1998 with low SAR, and high efficiency and various antenna diversity systems rated as the most efficient on the market, and he also holds more than 15 patents. He started the area of measurements of small active terminals including the antenna, and developed a measuring setup that now is used worldwide. He is involved in small terminals for 4G including several antennas (MIMO systems) to enhance the data communication.

David A. Sánchez-Hernández

David A. Sánchez-Hernández is with Universidad Politécnica de Cartagena, Spain, where he leads the microwave, radiocommunications, and electromagnetism engineering research group (GIMRE, http://www.gimre.upct.es). He has authored over 35 scientific papers, 75 conference contributions, and eight books, holds six patents, and also acts regularly as a reviewer of many IET and IEEE publications and conferences. He has also been awarded several national and international prices, including the R&D J. Langham Thompson Premium, awarded by the Institution of Electrical Engineers (formerly The Institution of Engineering and Technology, IET), and the i-patentes award by the Spanish Autonomous Region of Murcia to innovation and technology transfer excellence. His research interests encompass all aspects of the design and application of printed multiband antennas for mobile communications, electromagnetic dosimetry issues, and MIMO techniques.

Manos M. Tentzeris

Manos M. Tentzeris received a diploma degree in electrical and computer engineering from the National Technical University of Athens (Magna Cum Laude) in Greece and M.S. and Ph.D. degrees in electrical engineering and computer science from the University of Michigan, Ann Arbor, Michigan. He is currently an associate professor with the School of ECE, Georgia Tech, Atlanta, Geogia. He has published more than 280 papers in refereed journals and conference proceedings and two books and 10 book chapters. Dr. Tentzeris has helped develop academic programs in highly integrated/multilayer packaging for RF and wireless applications using ceramic and organic flexible materials, paper-based RFID's and sensors, microwave MEMs, SOP-integrated (UWB, multiband, conformal) antennas and adaptive numerical electromagnetics (FDTD, multiresolution algorithms), and he also heads the ATHENA research group (20 researchers). He is the Georgia Electronic Design Center associate director for RFID/sensors research, and he was the Georgia Tech NSF-Packaging Research Center associate director for RF research and the RF alliance leader from 2003 to 2006. He is also the leader of the RFID Research Group of the Georgia Electronic Design Center (GEDC) of the State of Georgia. He was the recipient/corecipient of the 2007 IEEE APS Symposium Best Student Paper Award, the 2007 IEEE IMS Third Best Student Paper Award, the 2007 ISAP 2007 Poster Presentation Award, the 2006 IEEE MTT Outstanding Young Engineer Award, the 2006 Asian-Pacific Microwave Conference Award, the 2004 IEEE Transactions on Advanced Packaging Commendable Paper Award, the 2003 NASA Godfrey "Art" Anzic Collaborative Distinguished Publication Award, the 2003 IBC International Educator of the Year Award, the 2003 IEEE CPMT Outstanding Young Engineer Award, the 2002 International Conference on Microwave and Millimeter-Wave Technology Best Paper Award (Beijing, China), the 2002 Georgia Tech-ECE Outstanding Junior Faculty Award, the 2001 ACES Conference Best Paper Award, the 2000 NSF CAREER Award, and the 1997 Best Paper Award of the International Hybrid Microelectronics and Packaging Society. He was also the 1999 Technical Program Co-Chair of the 54th ARFTG Conference, Atlanta, Georgia and the Chair of the 2005 IEEE CEM-TD

Workshop and he is the vice-chair of the RF Technical Committee (TC16) of the IEEE CPMT Society. He has organized various sessions and workshops on RF/ wireless packaging and integration, RFIDs, numerical techniques/wavelets, in IEEE ECTC, IMS, VTC, and APS Symposia, in all of which he is a member of the Technical Program Committee in the area of components and RF. He will be the TPC Chair for IEEE IMS 2008 Symposium. He is the associate editor of *IEEE Transactions on Microwave Theory and Techniques, IEEE Transactions on Advanced Packaging,* and *International Journal on Antennas and Propagation.* Dr. Tentzeris was a visiting professor with the Technical University of Munich, Germany, for the summer of 2002, where he introduced a course in the area of high-frequency packaging. He has given more than 50 invited talks in the same area to various universities and companies in Europe, Asia, and America. He is a senior member of the IEEE, a member of the URSI-Commission D, a member of MTT-15 committee, an associate member of the EuMA, a fellow of the Electro-magnetic Academy, and a member of the Technical Chamber of Greece.

Juan F. Valenzuela-Valdés

Juan F. Valenzuela-Valdés received a degree in telecommunications engineering from the Universidad de Malaga, Spain, in 2003. In 2004 he worked at CETE-COM (Malaga), and also joined the Department of Information Technologies and Communications, Technical University of Cartagena, Spain, where he is working towards his Ph.D. He is currently Head of Research at EMITE Ingenieria SLNE his current research areas cover MIMO communications and SAR measurements. He has two patents pending in these areas.

Index